# COLLECTED WORKS

OF

# COUNT RUMFORD

VOLUME III

# COLLECTED WORKS
## OF
# COUNT RUMFORD

EDITED BY SANBORN C. BROWN

VOLUME III

DEVICES AND TECHNIQUES

THE BELKNAP PRESS OF
HARVARD UNIVERSITY PRESS
CAMBRIDGE, MASSACHUSETTS
1969

# PREFACE

While Volume II contains many of Count Rumford's practical applications of his studies in heat, his writings on these topics were so extensive that they could not all be included in a single volume. Volume III, therefore, brings together a number of papers that Rumford wrote with the aim of increasing the efficiency of various industrial processes which required heat and for which the Count advocated the use of steam. Also included here are his papers on the use of steam heat for heating houses and institutions, and his very long essay on kitchen stoves and ovens as well as pots and pans.

Insofar as it has been practical, this edition of Rumford's Collected Works has reproduced his original figures to enhance the faithfulness of reproduction. A deviation from this procedure occurs in the long paper in this volume entitled "On the Construction of Kitchen Fire-places and Kitchen Utensils." In this paper Rumford himself changed his style of using figures. In almost all of his papers the figures are plates that are appended to the text. However, in this particular article he interspersed with the text 95 figures of various details of construction of his kitchen equipment. Many of these are simple line drawings and all of them seem to have been faithfully reproduced in the Academy's edition of *The Complete Works of Count Rumford*. Since it was determined that no new insights or more faithful reproduction would result from reproduction of the originals of these 95 figures, we have included here the relithographed figures of the previous edition of these *Works*.

The paper "On the Specific Gravity, Strength, Diameter, and Cohesion of Silk" was omitted from the Academy edition. This paper was sent to The Royal Society in June of 1786, but for some reason was never included in their *Proceedings*. Its existence has been known to scholars and Rumford himself refers to it a number of times in his publications. It was part of a larger work on the strength of materials which Rumford never finished, and it is probable that the failure of The Royal Society to publish it in their *Proceedings* was owing to their expectation of receiving a more complete experimental study before issuing it in published form.

By no means all of Rumford's work polarized around the study of heat and its practical applications, and two papers included in this volume illustrate his widely diverse interests. One covers his study of the advantage of using wheels with wide tires for travel on rough roads and the other is the plan for construction of a frigate, which was accepted by the British Admiralty but never put into effect.

Sanborn C. Brown

# CONTENTS

# COLLECTED WORKS

OF

# COUNT RUMFORD

VOLUME III

# OF THE USE OF STEAM

## AS A

## VEHICLE FOR TRANSPORTING HEAT.

MORE than fifty years ago, Colonel William Cook, in a paper presented by him to the Royal Society, and published in their Transactions, made a proposal for warming rooms by means of metallic tubes filled with steam, and communicated with a boiler situated out of the room; which proposal was accompanied by an engraving, which showed, in a manner perfectly clear and distinct, how this might be effected. Since that time this scheme has frequently been put in practice with success, both in this country and on the continent.*

Many attempts have been made, at different periods, to heat liquids by means of steam introduced into them; but most of these have failed; and, indeed, until it was known that fluids are non-conductors of heat, and, con-sequently, that heat cannot be made to *descend* in them (which is a recent discovery), these attempts could hardly succeed; for, in order to their being successful, it is absolutely necessary that the tube which conveys the hot steam should open into the *lowest part* of the vessel which contains the liquid to be heated, or nearly on a level with its bottom; but as long as the erroneous opinion

* Although one should naturally imagine that the notoriety of these facts would have been sufficient to prevent all attempts in our day to claim a right to this invention, yet it is said that a patent for it was taken out only a few years ago.

I

obtained, that heat could pass in fluids *in all directions*, there did not appear to be any reason for placing the opening of the steam-tube *at the bottom of the vessel*, while many were at hand which pointed out other places as being more convenient for it.

But to succeed in heating liquids by steam, it is necessary, not only that the steam should enter the liquid at the bottom of the vessel which contains it, but also that it should enter it *coming from above*.

The steam-tube should be in a vertical position, and the steam should descend through it previous to its entering the vessel, and mixing with the liquid which it is to heat; otherwise this liquid will be in danger of being forced back by this opening into the steam-boiler: for, as the hot steam is suddenly condensed on coming into contact with the cold liquid, a vacuum is necessarily formed in the end of the tube; into which vacuum the liquid in the vessel, pressed by the whole weight of the incumbent atmosphere, will rush with great force and with a loud noise; but if this tube be placed in a vertical position, and if it be made to rise to the height of six or seven feet above the level of the surface of the liquid which is to be heated, the portion of the liquid which is thus forced into the lower end of the tube will not have time to rise to that height before it will be met by steam, and obliged to return back into the vessel.

There will be no difficulty in arranging the apparatus in such a manner as effectually to prevent the liquid to be heated from being forced backwards into the steam-boiler; and when this is done, and some other necessary precautions to prevent accidents are taken, steam may be employed, with great advantage, for heating liquids, and for keeping them hot in a variety of cases in which fire, applied immediately to the bottoms of the containing vessels, is now used.

In dyeing, for instance, in bleaching, and in brewing, and in the processes of many other arts and manufactures, the adoption of this method of applying heat would be attended not only with a great saving of labour and of fuel, but also of a considerable saving of expense in the purchase and repairs of boilers, and of other expensive machinery: for, when steam is used instead of fire, for heating their contents, boilers may be made extremely thin and light; and as they may easily be supported and strengthened by hoops and braces of iron, and other cheap materials, they will cost but little, and seldom stand in need of repairs.

To these advantages we may add others of still greater importance. Boilers intended to be heated in this manner may, without the smallest difficulty, be placed in any part of a room, at any distance from the fire, and in situations in which they may be approached freely on every side. They may, moreover, easily be so surrounded with wood, or with other cheap substances which form warm covering, as most completely to confine the heat within them and prevent its escape. The tubes by which the steam is brought from the principal boiler (which tubes may conveniently be suspended just below the ceiling of the room) may, in like manner, be covered so as almost entirely to prevent all loss of heat by the surfaces of them, and this to whatever distances they may be made to extend.

In suspending these steam-tubes, care must, however, be taken to lay them in a situation *not perfectly horizontal,* under the ceiling, but to incline them at a small angle, making them rise gradually from their junction

with the top of a large vertical steam-tube, which con-
nects them with the steam-boiler, quite to their farthest
extremities; for, when these tubes are so placed, it is
evident that all the water formed in them, in consequence of the condensation of the steam in its passage
through them, will run backwards, and fall into the
boiler, instead of accumulating in them and obstructing
the passage of the steam (which it would not fail to do
were there any considerable bends or wavings, upwards
and downwards, in these tubes), or of running forward
and descending with the steam into the vessels containing the liquids to be heated, — which would happen if
these tubes inclined *downwards*, instead of inclining upwards, as they recede from the boiler.

In order that clear and distinct ideas may be formed
of the various parts of this apparatus, even without
figures, I shall distinguish each part of it by a specific
name. The vessel in which water is boiled in order to
generate steam — and which, in its construction, may be
made to resemble the boiler of a steam-engine — I shall
call the *steam-boiler;* the vertical tube which, rising
up from the top of the boiler, conveys the steam into
the tubes (nearly horizontal) which are suspended from
the ceiling of the room, I shall call the *steam-reservoir.*
To the horizontal tubes I shall give the name of *conductors of steam;* and to the (smaller) tubes which, descending perpendicularly from these *horizontal conductors*,
convey the steam to the liquids which are to be heated,
I shall exclusively appropriate the appellation of *steam-
tubes.*

The vessels in which the liquids that are to be heated
are put, I shall call the *containing vessels.* These vessels
may be made of any form; and, in many cases, they

may, without any inconvenience, be constructed of wood, or of other cheap materials, instead of being made of costly metals, by which means a very heavy expense may be avoided; or they may be merely pits sunk in the ground, and lined with stone or with bricks.

Each *steam-tube* must descend *perpendicularly* from the *horizontal conductor* with which it is connected, to the level of the bottom of the *containing vessel* to which it belongs; and, moreover, must be furnished with a good cock, perfectly steam-tight, which may best be placed at the height of about six feet above the level of the floor of the room.

This *steam-tube* may either descend *within the vessel* to which it belongs, or *on the outside of it*, as shall be found most convenient. If it comes down on the outside of the vessel, it must enter it at its bottom by a short horizontal bend; and its junction with the bottom of the vessel must be well secured, to prevent leakage. If it comes down into the vessel on the inside of it, it must descend to the bottom of it, or at least to within a very few inches of the bottom of it; otherwise the liquid in the vessel will not be uniformly or equally heated.

When the steam-tube is brought down on the inside of the containing vessel, it may either come down perpendicularly and without touching the sides of it, or it may come down on one side of the vessel and in contact with it.

When several steam-tubes belonging to different containing-vessels are connected with one and the same horizontal steam-conductor, the upper end of each of these tubes, instead of being simply attached by solder or by rivets to the under side of the conductor, must *enter* at least one inch *within the cavity of it;* otherwise

the water resulting from a condensation of a part of the
steam in the conductor by the cold air which surrounds
it, instead of finding its way back into the steam-boiler,
will descend through the steam-tubes, and mix with the
liquids in the vessels below; but when the open ends
of these tubes *project upwards within the steam-conductor*,
though it be but to a small height above the level of its
under side, it is evident that this accident cannot happen.

It is not necessary to observe here, that, in order that
the ends of the steam-tubes may project *within* the *horizontal conductor*, the diameters of the former must be
considerably less than the diameter of the latter.

To prevent the loss of heat arising from the cooling
of the different tubes through which the steam must
pass in coming from the boiler, all those tubes should
be well defended from the cold air of the atmosphere,
by means of warm covering; but this may easily be
done, and at a very trifling expense. The horizontal
conductors may be enclosed within square wooden tubes,
and surrounded on every side by charcoal dust, fine
sawdust, or even by wool; and the steam-tubes, as
well as the reservoir of steam, may be surrounded, first
by three or four coatings of strong paper, firmly attached
to them by paste or glue, and covered with a coating
of varnish, and then by a covering of thick coarse
cloth. It will likewise be advisable to cover the horizontal conductors with several coatings of paper; for, if
the paper be put on to them while it is wet with the
paste or glue, and if care be taken to put it on in long
slips or bands, wound regularly round the tube in a
spiral line from one end of it to the other, this covering will be useful, not only by confining more effectually
the heat, but also by adding very much to the strength

of the tube, and rendering it unnecessary to employ thick and strong sheets of metal in the construction of it.

However extraordinary and incredible it may appear, I can assert it as a fact, which I have proved by repeated experiments, that if a hollow tube, constructed of sheet copper $\frac{1}{20}$ of an inch in thickness, be covered by a coating only twice as thick, or $\frac{1}{10}$ of an inch in thickness, formed of layers of strong paper, firmly attached to it by good glue, the strength of the tube will be *more than doubled* by this covering.

I found by experiments, the most unexceptionable and decisive, — of which I intend at some future period to give to the public a full and detailed account, — that the strength of paper is such, when several sheets of it are firmly attached together with glue, that a solid cylinder of this substance, the transverse section of which should amount to only one superficial inch, would sustain a weight of 30,000 pounds avoirdupois, or above 13 tons, suspended to it, without being pulled asunder or broken.

The strength of hemp is still much greater, when it is pulled equally in the direction of the length of its fibres. I found, from the results of my experiments with this substance, that a cylinder of the size above mentioned, composed of the straight fibres of hemp glued together, would sustain 92,000 pounds without being pulled asunder.

A cylinder of equal dimensions, composed of the strongest iron I could ever meet with, would not sustain more than 66,000 pounds weight; and the iron must be very good not to be pulled asunder with a weight equal to 55,000 pounds avoirdupois.

I shall not, in this place, enlarge on the many advantages that may be derived from a knowledge of these curious facts. I have mentioned them now, in order that they may be known to the public; and that ingenious men, who have leisure for these researches, may be induced to turn their attention to a subject, not only very interesting on many accounts, but which promises to lead to most important improvements in mechanics.

I cannot return from this digression without just mentioning one or two results of my experimental investigations relative to the force of cohesion, or strength of bodies, which certainly are well calculated to excite the curiosity of men of science.

The strength of bodies of different sizes, *similar in form* and composed of the *same substance*, — or the forces by which they resist being pulled asunder by weight suspended to them, and acting in the direction of their lengths, — *is not in the simple ratio of the areas of their transverse sections*, or of their *fractures*, but in a higher ratio; and this ratio is different in different substances.

The *form* of a body has a considerable influence on its strength, *even when it is pulled in the direction of its length*.

All bodies, even the most brittle, appear to be *torn asunder*, or their particles separated, or fibres broken, *one after the other*; and hence it is evident that that *form* must be most favourable to the strength of any given body, pulled in the direction of its length, which enables the greatest number of its particles, or longitudinal fibres, to be separated to the greatest possible distance short of that at which the force of cohesion is overcome, before *any of them* have been forced *beyond* that limit.

It is more than probable that the apparent strength of different substances depends much more on the number of their particles that come into action before any of them are forced beyond the limits of the attraction of cohesion, than on any specific difference in the intensity of that force in those substances.

But to return to the subject more immediately under consideration. As it is essential that the steam employed in heating liquids, in the manner before described, should enter the containing vessel at or very near its bottom, it is evident that this steam must be sufficiently strong or elastic to overcome not only the pressure of the atmosphere, but also the additional pressure of the superincumbent liquid in the vessel; the steam-boiler must therefore be made strong enough to confine the steam, when its elasticity is so much increased, by means of additional heat, as to enable it to overcome that resistance. This increase of the elastic force of the steam need not, however, in any case, exceed a pressure of five or six pounds upon a square inch of the boiler, or *one third part*, or *one half*, of an atmosphere.

It is not necessary for me to observe here, that in this and also in all other cases where steam is used as a vehicle for conveying heat from one place to another, it is indispensably necessary to provide *safety-valves* of two kinds, — the one for letting a part of the steam escape, when, on the fire being suddenly increased, the steam becomes so strong as to expose the boiler to the danger of being burst by it; * the other for admitting

* The steam which escapes out of the boiler through the safety-valve may very easily be made to pass into the reservoir of water which feeds the boiler, and be condensed there; which will warm that water, and by that means save a quantity of heat which otherwise would escape into the atmosphere and be lost.

air into the boiler, when, in consequence of the diminution of the heat, the steam in the boiler is condensed,
and a vacuum is formed in it ; and when, without this
valve, there would be danger, either of the sides of the
boiler being crushed, and forced inwards by the pressure
of the atmosphere from without, or of the liquid in
the containing vessels being forced upwards into the
horizontal steam-conductors, and from thence into the
steam-boiler. The last-mentioned accident, however,
cannot happen, unless the cocks in some of the steam-
tubes are left open. The two valves effectually prevent all accidents.

The reader will, no doubt, be more disposed to pay
attention to what has here been advanced on this interesting subject, when he is informed that the proposed
scheme has already been executed on a very large scale,
and with complete success ; and that the above details
are little more than exact descriptions of what actually
exists.

A great mercantile and manufacturing house at Leeds,
that of Messrs. Gott and Company, had the courage, notwithstanding the mortifying prediction of all their
neighbours, and the ridicule with which the scheme was
attempted to be treated, to erect a *dyeing-house*, on a very
large scale indeed, on the principles here described and
recommended.

On my visit to Leeds in the summer of the year
1800, I waited on Mr. Gott, who was then mayor of
the town, and who received me with great politeness,
and showed me the cloth-halls and other curiosities of
the place; but nothing he showed me interested me
half so much as his own truly noble manufactory of
superfine woollen cloths.

I had seen few manufactories so extensive, and none so complete in all its parts. It was burnt to the ground the year before, and had just been rebuilt on a larger scale, and with great improvements in almost every one of its details.

The reader may easily conceive that I felt no small degree of satisfaction, on going into the dyeing-house, to find it fitted up on principles which I had some share in bringing into repute, and which Mr. Gott told me he had adopted in consequence of the information he had acquired in the perusal of my *Seventh* Essay.[1]

He assured me that the experiment had answered, even far beyond his most sanguine expectations; and, as a strong proof of the utility of the plan, he informed me that his next-door neighbour, who is a dyer by profession, and who at first was strongly prejudiced against these innovations, had adopted them, and is now convinced that they are real improvements.

Mr. Gott assured me that he had no doubt but they would be adopted by every dyer in Great Britain in the course of a very few years.

The dyeing-house of Messrs. Gott and Company, which is situated on the ground floor of the principal building of the manufactory, is very spacious, and contains a great number of coppers, of different sizes; and as these vessels, some of which are very large, are distributed about promiscuously, and apparently without any order in their arrangement, in two spacious rooms, — each copper appearing to be insulated, and to have no connection whatever with the others, — all of them together form a very singular appearance.

The rooms are paved with flat stones, and the brims of all the coppers, great and small, are placed at

the same height (about three feet) above the pavement.
Some of these coppers contain upwards of 1800 gallons;
and they are all heated by steam from *one steam-boiler*,
which is situated in a corner of one of the rooms,
almost out of sight.

The horizontal tubes, which serve to conduct the
steam from the boiler to the coppers, are suspended
just below the ceiling of the rooms: they are made,
some of lead and some of cast-iron, and are from
four to five inches in diameter; but when I saw them,
they were naked, or without any covering to confine
the heat. On my observing to Mr. Gott that coverings
for them would be useful, he told me that it was
intended that they should be covered, and that coverings
would be provided for them.

The vertical *steam-tubes*, by which the steam passes
down from the horizontal *steam-conductors* into the coppers, are all constructed of lead; and are from $\frac{3}{4}$ of an
inch to $2\frac{1}{2}$ inches in diameter, being made larger or
smaller according to the sizes of the coppers to which
they belong. These steam-tubes all pass down on the
*outsides* of their coppers, and enter them horizontally at
the level of their bottoms. Each copper is furnished
with a brass cock, for letting off its contents; and it is
filled with water from a cistern at a distance, which is
brought to it by a leaden pipe. The coppers are all
surrounded by thin circular brick walls, which serve not
only to support the coppers, but also to confine the
heat.

The rapidity with which these coppers are heated by
means of steam is truly astonishing. Mr. Gott assured
me that one of the largest of them, containing upwards
of 1800 gallons, when filled with cold water from the cis-

tern, requires no more than *half an hour* to heat it till it actually boils! By the greatest fire that could be made under such a copper, it would hardly be possible to make it boil in less than an hour.

It is easy to perceive that the *saving of time* which will result from the adoption of this new mode of applying heat will be very great; and it is likewise evident that it may be increased almost without limitation, merely by augmenting the diameter of the steam-tube. Care must, however, be taken, that the boiler be sufficiently large to furnish the quantities of steam required. The *saving of fuel* will also be very considerable. Mr. Gott informed me that, from the best calculation he had been able to make, it would amount to near two thirds of the quantity formerly expended, when each copper was heated by a separate fire.

But these savings are far from being the only advantages that will be derived from the introduction of these improvements in the management of heat. There is one, of great importance indeed, not yet mentioned, which alone would be sufficient to recommend the very general adoption of them. As the heat communicated by steam can never exceed the mean temperature of boiling water by more than a very few degrees, the substances exposed to it can never be injured by it.

In many arts and manufactures this circumstance will be productive of great advantages, but in none will its utility be more *apparent* than in cookery, and especially in public kitchens, where great quantities of food are prepared in large boilers; for, when the heat is conveyed in this manner, all the labour now employed in stirring about the contents of those boilers, to prevent the victuals from being spoiled by burning to the bottoms of them, will be unnecessary, and the loss of heat occasioned by this stirring prevented; and, instead of expensive coppers or metallic boilers, which are sometimes unwholesome, and always difficult to be kept clean, and often stand in need of repairs, common wooden tubs may, with great advantage, be used as culinary vessels; and their contents may be heated by *portable fireplaces*, by means of steam-boilers attached to them.

As these portable fireplaces and their steam-boilers may, without the smallest inconvenience, be made of such weight, form, and dimensions, as to be easily transported from one place to another by two men, and be carried through a doorway of the common width, with this machinery, and the steam-tubes belonging to it, and a few wooden tubs, a complete public kitchen, for supplying the poor and others with soups and also with puddings, vegetables, meat, and all other kinds of food prepared by *boiling*, might be established in half an hour in any room in which there is a chimney (by which the smoke from the portable fireplace can be carried off); and when the room should be no longer wanted as a kitchen, it might, in a few minutes, be cleared of all this culinary apparatus, and made ready to be used for any other purpose.

This method of conveying heat is peculiarly well adapted for heating baths. It is likewise highly probable that it would be found useful in the bleaching business and in washing linen. It would also be very useful in all cases where it is required to keep any liquid at about the boiling-point for a long time without making it boil; for the quantity of heat admitted may be very nicely regulated by means of the brass cock belonging to the steam-tube. Mr. Gott showed me a boiler in which shreds of skins were digesting in order to make glue, which was heated in this manner; and in which the heat was so regulated that, although the liquid never actually boiled, it always appeared to be upon the very point of beginning to boil.

This temperature had been found to be best calculated for making good glue. Had any other *lower* temperature been found to answer better, it might have been kept up with the same ease, and with equal precision, by regulating properly the quantity of steam admitted.

I need not say how much this country is obliged to Mr. Gott and his worthy colleagues. To the spirited exertions of such men, who abound in no other country, we owe one of the proudest distinctions of our national character, that of being *an enlightened and an enterprising people.*

In fitting up the great kitchen at the house of the Royal Institution, I availed myself of that opportunity to show, in a variety of different ways, how steam may be usefully employed in heating liquids.

On one side of the room, opposite to the fireplace, and where there is no appearance of any chimney, I fitted up a steam-boiler, of cast-iron, which, to confine the heat, is so completely covered up by the brickwork in which it is set, that no part of it is seen. This boiler is supplied with water from a reservoir at a distance (which is not seen), and by means of a cock, which is regulated by an hollow floating ball of thin copper, the water in the boiler always stands at the same height or level.

The steam from this boiler rises up perpendicularly in a tin tube, which is concealed in a square wooden tube, by the side of the wall of the room, and enters an horizontal tin tube (concealed in the same manner) which lies against the wall and just under the ceiling.

From this horizontal steam-*conductor* three tubes descend perpendicularly (concealed in three square wooden tubes), and enter three different kitchen boilers (on a level with their bottoms), which are set in brickwork against the same side of the room where the steam-boiler is situated.

As each of these boilers has its separate fireplace, properly furnished with a good double door and register ash-pit door, and also with a canal, furnished with a damper, for carrying off the smoke, either of these three boilers may be used for cooking, either with a fire made under it, or with steam brought into it from the neighbouring steam-boiler.

The object I had principally in view in this arrangement was to show, in the most striking and convincing manner, that all the different processes of cookery which are performed by boiling, such as boiling meat and vegetables *in boiling water*, making soups, stewing, etc., may in all cases be performed quite as well, and in many much better, by heating the liquid which is to be boiled, and keeping it boiling, by admitting hot steam *into it*, than by making a fire *under it*.

By using one of these boilers *alternately* in these two ways, on different days, in preparing the same kind of food, I concluded that all doubts on this subject would be most effectually removed.

To exhibit in a manner still more striking the application of steam to the boiling of liquids for culinary purposes, the following arrangement has been made and completed. A horizontal steam-conductor (concealed in a square wooden tube), communicating at right angles with the steam-conductor before described, passes, just below the ceiling, from the middle of one side of the room to the middle of the ceiling, and ends in a vessel in the form of a flat drum, about 10 inches in diameter and 5 inches high, which is attached to the ceiling perpendicularly over the centre of a large table which is placed in the middle of the room.

On the outside of this drum, or short hollow cylinder (which is made of tin and covered with wood, to confine the heat), there are, at equal distances, four projecting horizontal tubes, each about 1 inch in diameter and 2 inches long, which communicate with the inside of the drum. These tubes all point to the same centre, namely, to the centre of the drum.

To each of these short horizontal tubes there is fixed one end of a steam-tube composed of three pieces, fixed to each other, and movable, by means of joints, which are all steam-tight.

The end of this compound flexible steam-tube is united to the end of the short tube which projects from the side of the drum, by means of a steam-joint, in such a manner that the steam-tube attached to the drum, and communicating with it, may either be folded up in joints or lengths just under the ceiling, or it may be made to hang down from the end of the short tube to which it is attached. The lower joint, or rather division, of this flexible steam-tube, which reaches nearly to the top of the table, is furnished with a brass cock, by which it is occasionally closed, or, rather, by which it is always kept closed when it is not in actual use.

I might perhaps spare myself the trouble of describing the manner in which this culinary steam-apparatus is used, as the imagination of the reader will most probably have run before me. I shall, however, just mention a very striking and pleasing manner of making the experiment, in which the action of this machinery will be exhibited to great advantage.

If the cold water which is to be heated and made to boil by the steam is put into a large glass bowl or jar, on plunging the lower end of one of the flexible steam-tubes into the water, and then opening the steam-cock, the agitation into which the water in the glass vessel will be thrown will be visible through the glass; and the passage of the steam, in its elastic form, upwards through the water into the air, *after the water has become boiling hot* and not before, will be an instructive, as well as an amusing experiment.

Those of the flexible steam-tubes which are not in actual use are kept so folded up (in order to their being out of the way) that their two upper divisions, lying by the side of each other in a horizontal position, are just under the ceiling of the room; while their lower divisions hang vertically downwards, pointing towards the table.

In order that the kitchen may not be filled with steam when any of the boilers on the side of the room are used, their covers are all furnished with steam-tubes, which, communicating by a particular contrivance with a horizontal steam-tube which lies immediately over these boilers just under the ceiling, and which, by passing through the wall of the building, opens into the external air, all the waste steam from these boilers is carried out of the kitchen.

Before I conclude this Essay, I shall add a few observations concerning an application of steam which has not yet, to my knowledge, been made, but which there is much reason to think would turn out to be of very great importance indeed in many cases. This is the employing of it for communicating *degrees of heat above that of boiling water.*

I was led to meditate on this subject by an account I received, not long ago, of some very surprising effects which were produced in bleaching, by using the steam of a very strong solution of potash for boiling the linen, instead of water ; as I was confident that no part of the alkali could possibly be evaporated in this process, I could not account in any other way for the effects produced, but by supposing them to have been owing to the *high temperature* of the steam which rose from this strong lixivium; and as steam, at a high temperature, might easily be procured and applied to the linen without the use of the alkali, I thought it would be worth while to try the experiment with hot steam produced from pure water. I mentioned this idea to Mr. Duffin, Secretary of the Linen Board in Ireland, who is himself concerned, in an extensive way, in the bleaching business, who has promised to make some experiments on this subject, which I took the liberty to point out and to recommend to him as being likely to lead to interesting results.

Meditating on the various uses to which *hot* or (which is the same thing) *strong steam* might be applied, it occurred to me that it would probably be found to be extremely useful in *alum works,* for concentrating the liquor from which alum is crystallized. There are, as is well known, many difficulties attending the evaporation and concentration of that liquid; and it is never done without occasioning a very considerable expense,

as well for fuel, of which large quantities are consumed,
as also on account of the frequent repairs of the pans,
which are found to be necessary.

Most, if not all these difficulties might, I think, be
avoided by introducing strong steam into this liquor,
instead of concentrating it over a fire.  This concentration might certainly be effected as well, and probably
better and more expeditiously, by using hot steam, than
by the immediate use of the heat of a fire, and the
expense occasioned by the wear and tear of the apparatus
would, no doubt, be much less in the former case than
in the latter; and if it should be found (which is not
unlikely) that *some certain temperature* is more advantageous in this process than any other, *that temperature*,
when once discovered, may be preserved, with very little
variation, when steam is used (by placing a valve, loaded
with a proper weight, in the steam-tube, and obliging the
steam to lift that valve, in order to pass through the tube);
but there is no possibility of regulating, with any precision, the degrees of heat employed when liquids are
evaporated in boilers over a fire.

I would just point out one more application of steam,
which, if I am not much mistaken, will turn out to be
very advantageous indeed in many respects ; — it may be
employed in heating the fermented liquor from which
ardent spirits are distilled.

A proposal for introducing watery vapour into a
liquor from which pure ardent spirits are to be distilled,
or forced away by heat, will, no doubt, be thought very
extraordinary by those who have never meditated on
the subject; but when they shall have considered it with
attention, they will find reason to conclude that this
method of distilling bids fair to be very useful.  The

saving of expense for coppers and other costly utensils and machinery would be very considerable, and the danger of the flavour of the spirits being injured by the burning of the liquor to the sides of the copper would be entirely removed.

Steam has already been introduced, in several great manufactories in this country, into *drying-houses*, and employed with the best effects for heating and drying linen, cotton, and woollen goods, after they have been washed; it has also been used in the *drying-rooms* of several paper-manufactories. When it is used for any of these purposes, it should be introduced into tubes of large diameter, or into several smaller tubes, constructed of very thin sheet copper (or into any other metallic tubes, *having a large surface*, that would be cheaper); and these tubes should be placed nearly in a horizontal position in the *lower part* of the drying-room and *under* the goods that are to be dried; and (in order to economize the heat as much as possible) the water resulting from the condensation of the steam in the steam-tubes should be conducted by small tubes, well covered with warm covering, into the reservoir which feeds the steam-boiler.

# NOTE ON THE USE OF STEAM HEAT.

SEVERAL individuals with whom I have not the honour of being personally acquainted have applied to me within a short time for information with regard to the history of the use of the vapour of boiling water as a vehicle for conveying heat in the distillation of brandies, — a process which I have recommended in my Fifteenth Essay,[2] published at London in the month of May, 1802, and deposited the same month in the library of the Institute. Judging, from the extreme eagerness which they have manifested to obtain this information, and to have it in writing, that it is a question of establishing certain facts which are held to be important, I have thought it proper to give the Class information in this matter.

It is not so much to claim the advantage of having been the first to propose a useful process, and to teach the means of assuring its success, as to avoid being drawn into any sort of discussion in the matter, that I have decided to address myself to the Class on this occasion instead of furnishing the information in question to an individual. Foreseeing, moreover, that the Class might be called upon to give an opinion in this matter, I take the liberty of submitting to it a translation of certain paragraphs from my Fifteenth Essay.

WHEN the hall which it is desired to heat is very large, and has several large windows, it is indispensably necessary to begin by making the windows *double;* for without this precaution the continual cooling which will take place through single windows will be so great that, no matter how much wood is burned, it will never be possible to warm the apartment uniformly throughout, and as soon as the fire ceases to burn the room will quickly become cold.

There would be no use in employing the best stoves to remedy these inconveniences. Close to the stoves it will indeed be possible to feel the heat caused by their calorific radiations; but nothing can hinder the currents of cold air, caused by the cooling which takes place through the panes of glass, from spreading over the entire extent of the room.

Those particles of air in the room which are in immediate contact with the glass, finding themselves specifically heavier on account of this change of temperature, must necessarily descend and spread themselves over the pavement, forming currents which are perceptibly cold, and no doubt very injurious to health. But, when the windows are double, the layer of air which is enclosed between the two windows being an excellent non-conductor of heat, the inside window is well protected from cold from without; and, the descending currents of cold air just mentioned no longer existing, it would be easy, with good stoves moderately heated, to establish a pleasant and equable temperature, and to make it permanent, at a small expense.

By doubling the windows of the hall of the Institute which it is proposed to heat, it would be possible easily, and without much expense, to obtain a very important advantage besides that of which we have just spoken.

Since the hall is surrounded by very high buildings which are close to it, there is a deficiency of light in the hall which is very noticeable, especially in cloudy weather and towards the end of the day. By making the windows double, and using panes of ground glass for the outside windows, the amount of light in the hall would be much increased, and the light will be more equable, softer, and more agreeable.

As to the means of heating, it is certain, from the results of several decisive experiments, that steam stoves are preferable to every other sort, especially for large apartments.

1st. The heat which these stoves distribute in a room is singularly soft and agreeable, and never causes headache, as iron stoves do which are heated directly by the burning fuel.

2d. The temperature of a room warmed by steam can be regulated at pleasure with the greatest ease by means of a simple cock to close more or less the tube which conducts the steam from the boiler into the stove.

3d. As the boiler can without any inconvenience be placed outside of the hall, and even at a considerable distance, it may be put in an out-of-the-way place, where there will be every security against accidents from fire, and at the same time great ease in storing the wood intended for the boiler, and in regulating its consumption. It is necessary, however, to take care that the boiler be placed lower than the stove, in order that the water resulting from the condensation of steam in the stove may return to the boiler.

4th. Since the boiler will be provided with safety-valves, the stove will never be in danger either of being burst by the elastic force of the steam, or of being crushed by the pressure of the atmosphere; and on this account it may be constructed without difficulty of very thin sheets of copper, so that the expense of its construction ought not to be very great.

5th. These stoves may be made of any desired form; but the best shape is that of a cylindrical tube, or of a column, for this is the form which gives them the greatest strength to resist, without change of shape, the expansive force of the steam within and the pressure of the atmosphere on the outside.

6th. The steam should be introduced into the stove at its upper extremity; and in the lowest part of the stove there should be a tube to conduct into the boiler the water which results from the condensation of steam in the stove. In order that the tube which conducts the steam into the stove may not be visible in the apartment, it may be made to enter through the bottom of the stove, and then ascend inside, to within 2 or 3 inches of the upper end, where there should be an opening. As the vapour of boiling water is specifically lighter than atmospheric air, by bringing the steam into the upper part of the stove it presses upon the air in the stove, and drives it out by one of the safety-valves without mixing with it, so that this air is driven out quietly, and without first being warmed at the expense of the heat of the apparatus. This air must descend by the tube which serves to conduct the water from the stove into the boiler; and the valve by which it escapes into the atmosphere, being situated near the boiler, may open into a canal or a tube communicating with the chimney of the boiler fire-place. Then if, by the carelessness of the person having charge of the stove, there is too much steam, since it will follow the same road, it will escape by the chimney without diffusing itself into the apartment.

7th. The tube which carries the water resulting from the condensation of the steam in the stove back into the boiler must pass through the walls or cover of the boiler, and descend within it nearly to the bottom; and the extremity, being always beneath the water in the boiler, should be bent and turned upwards. All these precautions are necessary to prevent the steam in the boiler from ever finding its way into this tube.

8th. The steam-tube which communicates with the highest part of the stove should start from the highest part of the boiler, and this tube, as well as that which carries the water back from the stove to the boiler, should be well surrounded by suitable coverings, in order to preserve their heat. The boiler should also be well covered above and on every side, so as to protect it from the cold.

9th. Although the expenditure of water in this apparatus is almost nothing when the fire is properly regulated, so that when the boiler has been filled at the beginning of the autumn there is no need of touching it during the winter, or indeed for several years, — nevertheless, as it might easily happen that the fire should be driven too much, owing to carelessness, from time to time, so as to drive out part of the water in the form of steam by the safety-valve, it will be prudent to put a small reservoir of water near the boiler, and connected with it, so that one can readily examine it, and fill it as often as it shall prove necessary.

10th. The stove should be made of thin sheets of brass, and well soldered or brazed throughout in order to prevent the steam from forcing its way into the room; but great care must be taken not to leave the stove its metallic lustre on the outside. On the contrary, it must be painted on the outside, in order that it may diffuse more heat into the apartment. It is possible to give it the appearance of a marble or granite column, or to paint it in any other way which corresponds best on the outside with the furniture of the room. For the hall of the Institute I should propose to take away three of the wooden columns which are now there, and which do not support any thing, and to replace them by three copper columns of the same shape and size, and painted on the outside of the same colour. These three copper columns will be three steam stoves connected with a single boiler, which may be put in a little room on the ground floor, which happens to have a chimney, and which is used at present as a sort of lumber-room where articles of small value are stored.

In this way the hall of the Institute will be neither encumbered nor disfigured by the apparatus used for heating it in winter; and, being provided with double windows of ground glass, it will be lighter and more cheerful, and at the same time more quiet, being shut off from the cheerless and disagreeable objects which surround it on every side.

I shall say nothing of the advantage which would be

gained by the public from the introduction of a method
of heating which offers so many advantages on the
score both of elegance and of economy.

# DESCRIPTION OF A NEW BOILER,

CONSTRUCTED

## WITH A VIEW TO THE SAVING OF FUEL.

IT is well known that much is gained in the saving of fuel, when an extensive surface is given to that part of the boiler against which the flame strikes; but this advantage is often counterbalanced by great inconveniences. For a boiler of the form usually employed, having the bottom very much extended in proportion to its capacity, must necessarily present a great surface to the atmosphere, and the loss of heat, occasioned by the cold air coming in contact with this surface, may be more than sufficient to compensate the advantage derived from the extended surface of the bottom. And where the boiler is employed for producing steam, as it is indispensably necessary that it should be of a thickness sufficient to resist the expansive force of the steam, it is evident that, if the diameter be augmented (with a view to increase the surface of the bottom), a considerable expense is incurred on account of the additional strength that must be given to the sides.

Having been engaged in the year 1796 in a set of experiments in which I employed the steam of boiling water as a vehicle of heat, I had a boiler made for this purpose, on a new construction, which answered well, and even beyond my expectations; and as this boiler might be used with advantage in many cases, even where

29

it is only required to heat liquids in an open boiler, this, and another motive, which it would be useless to mention in this place, have lately induced me to construct one here (at Paris) and to present it to the Institute.

The object chiefly had in view in the construction of this boiler was to give it such a form, that the surface exposed to the fire should be great in comparison with its diameter and capacity ; and this without having a great surface exposed to the cold air of the atmosphere.

The body of the boiler is in the shape of a drum.   It is a vertical cylinder of copper 12 inches in diameter and 12 inches high, closed at top and at bottom by circular plates.

In the centre of the upper plate there is a cylindrical neck 6 inches in diameter and 3 inches high, shut at top by a plate of copper 3 inches in diameter and 3 lines in thickness, fastened down by screws.

This last plate is pierced by three holes, each about 5 lines in diameter.   The first, which is in the centre of the plate, receives a vertical tube, which conveys water to the boiler from a reservoir, which is placed above.   This tube, which descends in the inside of the boiler to within an inch above the circular plate which forms its bottom, has a cock near its lower end.   This cock is alternately opened and shut, by means of a floater which swims on the surface of the water contained in the body of the boiler.

The second of the holes in the 'plate that closes the neck of the boiler receives the lower end of another vertical tube, which serves to convey the steam from the boiler to the place where it is to be used.

The third hole is occupied by a safety-valve.

This description shows that there is nothing new in the construction or arrangement of the upper part of this boiler. In its lower part there is a contrivance for increasing its surface, which has been found very useful.

The flat circular bottom of the body of the boiler, which, as I said before, is 12 inches in diameter, being pierced by seven holes, each 3 inches in diameter, seven cylindrical tubes of thin sheet-copper, 3 inches in diameter and 9 inches long, closed below by circular plates, are fixed in these holes, and firmly riveted, and then soldered to the flat bottom of the boiler.

On opening the communication between the boiler and its reservoir, the water first fills the seven tubes, and then rises to the cylindrical body of the boiler; but it can never rise above 6 inches in the body of the boiler, for when it has got to that height, the floater is lifted to the height necessary for shutting the cock that admits the water.

When the height of the water in the boiler is diminished a few lines by the evaporation, the floater descends a little, the cock is again opened, and the water flows in again from the reservoir.

As the seven tubes that descend from the flat bottom of the body of this boiler into the fireplace are surrounded on all sides by the flame, the liquid contained in the boiler is heated, and made to boil in a short time, and with the consumption of a relatively small quantity of fuel; and when the vertical sides of the body of the boiler and its upper part are suitably enveloped, in order to prevent the loss of heat by these surfaces, this apparatus may be employed with much advantage in all cases where it is required to boil water for procuring steam.

And as in the case where the boiler is constructed on a great scale, the seven tubes that descend from the bottom of the boiler into the fire may be made of cast-iron, whilst the body of the boiler is composed of sheet-iron or sheet-copper, it is certain that a boiler of this kind, sufficiently large for a steam-engine, a dyeing-house, or a spirit-distillery, would cost much less than a boiler of the usual form, of equal surface and power.

But in all cases where it is required to produce a great quantity of steam, it will be always preferable to employ several boilers of a middling size, placed beside each other, and heated each by a separate fire, instead of using one large boiler heated by one fire.

I have shown in my Sixth Essay, on the management of fire and the economy of fuel, that beyond a certain limit there is no advantage derived from augmenting the capacity of a boiler.

It will be perceived that the boiler which I have the honour of presenting to this Society is of a form fit for being placed in a portative furnace, and it was actually intended for that purpose.

Its furnace, which is made of bricks, with a circular iron grate of 6 inches in diameter, is built in the inside of a cylinder of sheet-iron, 17 inches in diameter and 3 feet high, and can be easily transported from place to place by two men.

This cylinder of sheet-iron, which is divided into two parts, in order to facilitate the construction of the masonry, weighs only forty-six pounds. The masonry weighs about a hundred and fifty pounds, and the boiler twenty-two pounds.

In order to form an estimate of the advantage which the particular form of this boiler gives it in accelerating its heating, we may compare the extent of surface that it presents to the action of the fire with that of the flat bottom of a common boiler.

The diameter of the bottom of a cylindrical boiler being 12 inches, the surface is 113.88 square inches; but the surface of the sides of the seven tubes that descend from the flat bottom of our boiler (which is likewise 12 inches in diameter) is 593.76 square inches. Therefore the new boiler has a surface exposed to the direct action of the fire, more than five times greater than that of a boiler of equal diameter and of the ordinary form; how much this difference must affect the celerity of heating is easy to conceive.

In the manner in which boilers are usually set, their vertical sides are but little struck by the flame, and on that account I have not taken the effect of the sides into consideration in my estimate; but even taking them into account, the new boiler will always have a surface exposed to the fire at least twice as great as that of a common cylindrical boiler of the same diameter, as can easily be shown.

The new boiler being 12 inches in diameter and 12 inches high, and each of its seven tubes being 3 inches in diameter and 9 inches high, its surface is 1160.44 square inches, without reckoning the circular plate that closes its top, nor its neck.

The surface of the bottom and sides of a cylindrical boiler of 12 inches in diameter and 12 inches high will be 566.68 square inches.

As the quantity of heat that enters a boiler in a given time is in proportion to the extent of surface that the boiler presents to the fire, it is evident that, other circumstances being the same, a boiler with tubes descending from its bottom will be heated at least twice as soon as a cylindrical boiler of the same diameter with a flat bottom.

In order that a cylindrical boiler with flat bottom, surrounded by flame on all sides, might have the same extent of surface exposed to the fire as a boiler with tubes, it would be necessary to give it a diameter greater than that of the boiler with tubes in the proportion of the square root of 1160.44 to the square root of 566.68, that is, of 17.171 to 12.

Therefore, in order that a cylindrical boiler with a flat bottom might have the same extent of surface exposed to the fire as our boiler with tubes of 12 inches in diameter, it would be necessary to give it a diameter of 17.171 inches.

But if the diameter of a boiler intended for producing steam be increased, it is necessary, at the same time, to increase its thickness, in order to increase its strength.

The necessary increase of thickness, and the expense that it will occasion, can be easily calculated.

The effort that an elastic fluid exerts against the sides of the containing vessel is in proportion to the surface of a longitudinal and central section of the vessel, and consequently in proportion to the square of its diameter, the form remaining the same. Hence we may conclude, that a steam-boiler of a cylindrical form with a flat bottom, which has the same extent of surface exposed to the fire as a boiler of 12 inches in diameter with tubes, should be at least twice as thick as this last, in order to have an equal degree of strength for resisting the expansive power of the steam.

The boiler which I have the honour of presenting to

the Society is particularly intended to serve as a steam-boiler, but it may undoubtedly be applied to other purposes. Having shown it to M. Auzilly, son of a considerable soap-manufacturer of Marseilles, he thought that it might be employed with advantage in the making of soap ; and from what he told me of the process, and of the boilers employed in that art, I am persuaded that the experiment would succeed perfectly.

But, after all, it remains to be determined whether it would not be still more advantageous to employ steam as a vehicle of heat in the making of soap, instead of lighting the fire under the bottom of the vessel in which the soap is made.

The result of an experiment which we are to make, M. Auzilly and myself, will probably throw some light upon this question.

# EXPERIMENT

## USE OF THE HEAT OF STEAM, IN PLACE OF THAT OF AN OPEN FIRE, IN THE MAKING OF SOAP.

I HAD the honour of announcing to this Assembly, at the last meeting but one, that M. Auzilly and myself were to make an experiment on the use of steam in the making of soap. This experiment we have made, and with perfect success.

I have the honour to lay before the Society a piece of soap of about ten cubic inches, made in my laboratory by this new process, which required only six hours of boiling, whereas sixty hours and more are necessary in the ordinary method of making soap.

From all the appearances that we observed in the course of this experiment, and from its results, we think ourselves authorized to conclude that this new method of making soap cannot fail to be advantageous in every respect, and that it will soon be generally adopted.

We propose to repeat the experiment on a larger scale, as soon as we shall be able to procure the necessary utensils, and we beg the Society to appoint commissioners to be present during its execution.

As I intend to communicate to the Institute, upon a future occasion, all the details of our experiment, with an account of the apparatus we employed in it, I shall for the present make only one observation on the prob-able cause of the acceleration of the formation of soap, which we observed. I believe that this acceleration is due in great measure, if not entirely, to a motion of a peculiar kind in the mixture of oil and lye, occasioned by the sudden condensation of the steam introduced into the liquor. It is a sharp stroke, like that of a hammer, which made the whole apparatus tremble.

These strokes, which succeeded rapidly in certain cir-cumstances, and which were violent enough to be heard at a considerable distance, must necessarily have forced the particles of oil and alkali to approach each other, and consequently to unite.

As the violence of these strokes diminished greatly as soon as the liquid had acquired nearly the temperature of the steam, I propose to supply this defect by a par-ticular arrangement of the apparatus in the experiment we are going to make. I shall divide the vessel into two parts, by a horizontal diaphragm of thin sheet cop-per, and, causing a slow current of cold water to pass through the lower division or compartment of the ves-sel, I shall introduce steam into it, through a particular tube destined for that purpose, as soon as the mixture of oil and alkali which occupies the upper division of the vessel is become too hot for condensing the steam.

The steam which enters the water (always kept cold) that fills the lower compartment of the vessel will be condensed suddenly, and the sharp strokes which result will be communicated through the thin diaphragm to the hot liquid contained in the upper division of the vessel, and will, I expect, accelerate the union of the oil with the alkali. I shall then shut almost entirely the cock which admits steam into the upper division of the vessel, in order to prevent a useless consumption of steam and heat.

I shall not fail to give an account of the results of this new experiment to this Assembly ; and I shall rejoice if by any researches I shall be so happy as to contribute to the improvement of an art which is undoubtedly of great importance to society.

# SUPPLEMENTARY OBSERVATIONS

RELATING TO

## THE MANAGEMENT

OF

# FIRES IN CLOSED FIRE-PLACES.

# OF THE MANAGEMENT OF FIRES IN
## CLOSED FIRE–PLACES.

---

*Necessity of keeping the Doors of closed Fire-places well closed, and of regulating the Air that is admitted into them. — Account of some Experiments which showed in a striking Manner the very great Importance of those Precautions.—A Method is proposed for preventing the Passage of cold Air into the large Fireplaces of Brewhouse Boilers, Distillers' Coppers, Steam-Engine Boilers, etc., while they are feeding with Coals. — Bad Consequences which result from overloading closed Fire-places with Fuel. — Computations which show in a striking Manner the vast Advantages that will be derived from the Use of proper Care and Attention in the Management of Fire, and in the Direction and Economy of the Heat which results from the Combustion of Fuel.*

THOUGH I have already mentioned, more than once, the necessity of preventing the entrance of air into a closed fire-place by any other passage than by the register of the ash-pit door, and have strongly recommended the keeping of the door of the fire-place constantly closed ; yet, as I have since found that those precautions are even of more importance than I had imagined, I conceived that it might be useful to mention the subject again, and give an account of the series of experiments from the results of which I have acquired new light in respect to it.

In fitting up a large shallow circular kitchen boiler (one of those I put up in the kitchen of the house formerly occupied by the Board of Agriculture), I made an experiment which, though it appeared to me at the time to have succeeded perfectly, led me into an error that afterwards caused me a great deal of embarrassment. I constructed the fire-place of the boiler of a peculiar form for the express purpose of *burning the smoke;* imagining that if I could succeed in that attempt I should not only get more heat from any given quantity of coals, but also that the narrow horizontal canal that carried off the smoke from the fire-place to the chimney would be much less liable to be choked up by soot or dust. The fire-place was made rather longer than usual; and near the farther end of it there was a thin piece of fire-stone, placed edgewise, which run quite across it from side to side, a space being left about $2\frac{1}{2}$ inches wide between the lower edge of this stone and the bars of the grate, while the bottom of the boiler reposed on its upper edge.

From this description it is evident that the flame of the burning fuel, after rising up and striking against that part of the bottom of the boiler which was situated over the hither part of the fire-place, must necessarily pass under the lower edge of the stone just mentioned, in order to get into the canal leading to the chimney; and I fancied that, by taking care to keep that *narrow passage* constantly occupied by red-hot coals, the smoke being forced to pass through between them would necessarily take fire and burn. This actually happened; and, when I left a small opening in the door of the fire-place to give admittance to a little fresh air to facilitate and excite the combustion, the flame became so exceedingly vivid and clear that I promised myself great advantages from this new arrangement.

Being soon after engaged in putting up a large square boiler in the kitchen of the Foundling Hospital, I there introduced the same contrivance; but how great was my surprise on finding that, notwithstanding the extreme vivacity of the fire, the contents of the boiler could not be brought to boil in less time than five hours! The fire-place, it is true, was small, and the brick-work was new and wet; but I found that the quantity of coals consumed was such that, had there been no essential fault in the construction of the fire-place, nor in the management of the fire, the contents of the boiler ought, notwithstanding these unfavourable circumstances, to have boiled in less than one third part of the time that had been found necessary to bring it into a state of ebullition.

Having wasted two or three days in attempting to remedy the defects of this fire-place, without changing entirely the principles of its construction; concealing my disappointment from those who it was necessary should have confidence in my skill, by representing to them all that had been done as being a mere experiment, I pulled down the work to the foundation, and caused it to be rebuilt on principles which I knew could not fail to succeed, and which did succeed to the utmost of my expectations.

Though I ruminated often on this disappointment, I did not find out the real cause of my ill success for some months. This discovery was, however, at length made, and in such a manner as to leave no room for doubt.

Having, as an experiment, constructed in the kitchen of the Military Academy at Munich an apparatus for the performance of all the different processes of cookery, and to serve occasionally for warming a room with one and the same fire, thinking that the principles of the invention might be employed with advantage in the construction of cottage fire-places, on my return to this country I made the experiment at my lodgings in Brompton Row, Knightsbridge; and, desirous of accommodating the contrivance to what I think may be called a prejudice of Englishmen, I contrived the machinery in such a manner as to render the fire *visible*.

A small low grate was fixed in the middle of a large open kitchen fire-place, and on each side of it were fixed in brick-work two Dutch ovens, one above the other, the bottom of the lower oven on each side being nearly on a level with the top of the grate; and, as each of the ovens was surrounded by flues, I had hopes that by causing the flame and smoke of the open fire to incline downwards and enter a horizontal canal, situated just behind the fire, and there to separate to the right and left and circulate under the iron bottoms of the ovens, they would by that means be sufficiently heated to bake or to boil; and, even if the two upper ovens should not be found to be sufficiently heated to perform those processes of cookery, I thought, by leaving their doors open, they might at least be very useful, occasionally for warming the room, acting in the manner of a German stove. But the experiment was far from succeeding as I expected.

The current of flame and smoke which arose from the open fire was, without difficulty, made to bend its course downwards into the canal destined to receive it, and to circulate in the flues of the ovens; but, to my astonishment, I found that the ovens, instead of being heated, were barely warmed. An accident, however, very fortunately for me, discovered to me the real cause of the ill success of the experiment. Throwing a piece of paper on the top of the coals that were burning in the grate, in order to see if *the whole* of the large flame which I knew the paper must produce would be drawn downwards into the horizontal opening of the canal, situated behind the back of the grate, I was surprised to find that this flame was not only drawn into this opening, but that it appeared to be violently *driven downwards* to the very bottom of the canal.

In short, every appearance indicated that there was a very strong vertical *wind* that was continually blowing *directly downwards* into the opening of the canal; and it immediately occurred to me that, as this wind consisted of a stream of cold air, this air must necessarily cool the ovens almost as fast as the flame heated them; and I was no longer surprised at the ill success of my experiment.

On considering the subject with attention, I saw how impossible it must be for the current of hot vapour, flame, and smoke that rises from burning fuel, to be made to pass off *horizontally*, or to deflect considerably from its direct ascension *in contact with the cold air of the atmosphere*, without drawing after it a great deal of that cold air; and I now saw plainly why so much time and fuel were required to heat the boiler in the kitchen of the Foundling Hospital, in the experiments that were made with its first fire-place.

The cold air which entered the fire-place at its door,

and passing *over* the surface of the burning fuel entered the flues of the boiler with the flame, cooled the bottom of the boiler almost as fast as the flame heated it.

The waste of heat that is occasioned *precisely in this manner* in the fire-places of steam-engines, brewers' coppers, distillers' coppers, etc., must be very great indeed. To be convinced of this fact, nothing more is necessary than to see how very imperfectly the entrance into one of these fire-places is closed by its single door, ill fitted to its frame; what a length of time the door is left *wide open* while the fire is stirring or fresh coals are putting into the fire-place; and what an impetuous torrent of cold air rushes into the fire-place on those occasions.

As the cold air that comes into the fire-place in this manner, and passes *over* the burning coals, has very little to do in promoting the combustion of the fuel, and must necessarily be heated very hot in passing through the fire-place and through the whole length of the flues of the boiler, it is easy to see what an immense quantity of heat this air must steal and carry off into the atmosphere in its escape up the chimney.

To remedy this evil, the doors of all closed fire-places should be double, and they should be fitted to their frames with the greatest nicety, which may easily be done by making them shut against the front edge of their frames, instead of being fitted *into them* or into grooves made to receive them; and, when the fire is burning, these doors should be opened as seldom as possible and for as short a time as possible. I have already mentioned the necessity of these precautions in my sixth Essay,[3] but they are of so much importance that they can hardly be too often recommended, nor can too much pains be taken to show why they are so necessary.

In all cases where a fire-place is very large, and where, in consequence of the large quantity of coals consumed in it, the fire-place door is necessarily kept open a great deal, I would earnestly recommend the adoption of a contrivance which I think could not fail to turn out a complete remedy for the evil we have been describing; viz., the entrance of a torrent of cold air into the fire-place through its door-way.

The contrivance is this: to construct the floor or pavement of the area before the fire-place door in such a manner as to cut off all direct communication, without the fire-place in front of it, between the ash-pit and the fire-place door-way; and, when this is done, to build a porch, well closed above and on every side, immediately before the fire-place door, and in such a manner that the fire-place door may open into it.

This porch must have a door belonging to it, situated on the side opposite to the fire-place door, which door (that belonging to the porch) must open outwards, and must fit its door-frame with considerable nicety. There must also be a glass window either in this door or over it, or on one side of it, or in one of the side walls of the porch; and there must be sufficient room in the porch to allow of a certain provision of coals being lodged there and kept ready for use.

When fresh coals are to be thrown into the fire-place (as also when the door of the fire-place is to be opened for the purpose of stirring the fire, or for any other purpose), the person who is charged with the care of the fire enters the porch, and then, carefully shutting the door of the porch after him, he opens the fire-place door.

As no air can get into the porch from without, its door being closed, none can pass through it into the fire-place, and the fire-place door may be left open without the smallest inconvenience; and the person who tends the fire may take up as much time as he pleases in stirring it or feeding it with fresh fuel, for little or no derangement of the fire or loss of heat will result from these operations. The fire will continue to burn nearly in the same manner as it did before the fire-place door was opened; and those immense clouds of dense smoke which, to the annoyance of the whole neighbourhood, are now thrown out of the chimneys of all great breweries, distilleries, steam-engines, etc., as often as they are fed with fresh coals, will no longer make their appearance.

When these operations are finished, and the fire-place door is again closed, the door of the porch may be opened, and the provision of coals kept in the porch for immediate use may be again completed.

If the flame from the fire-place should be found to have any tendency to come into the porch, this may be easily checked by leaving a very small hole in the door of the porch for the admission of a small quantity of air, just enough to prevent this accident. This small hole might be furnished with a register.

But it is not merely through the opening by which the fuel is introduced that cold air furtively finds its way into closed fire-places. It frequently enters in much too large quantities by the ash-pit door-way, and, rushing up between the bars of the grate and mixing with the flame, serves to diminish instead of increasing the heat applied to the bottom of the boiler; and this never fails to happen when a *small fire is made in a large fire-place*, or when a part of the grate happens not to be covered with burning fuel, especially when there is no register to the ash-pit door.

It should be remembered that whenever more air enters a closed fire-place than is actually *decomposed* by the burning fuel, all that superabundant air not only is of no service whatever, but being itself heated at the expense of the fire, and going off hot by the chimney, occasions the loss of a quantity of heat that might have been usefully employed.

Ash-pit doors should always be furnished with registers of whatever size the fire-place may be, for they are always indispensably necessary to the good management of a fire; and, where small fires are occasionally made in large closed fire-places, the ascent of air through that part of the grate that is not covered with burning fuel should be prevented by sliding an iron plate under the bars of the grate, or by some other contrivance equally effectual.

If the closed fire-places of boilers, great and small, were properly constructed, and if due care were taken to introduce in a proper manner and to regulate the quantity of the air that is necessary to the perfect combustion of the fuel, their grates might be made considerably narrower than they now are, and the bottoms of their boilers might be placed at a greater height above them, from which arrangement several advantages would be derived; but as long as so little care is taken to keep the door of the fire-place well closed, and to prevent too much air from coming up through the grate by the openings between its bars, the bottom of the boiler must be placed very near the surface of the burning coals, otherwise so much more cold air than is wanted will find its way into the fire-place and mix with the flame that the bottom of the boiler cannot fail to be sensibly cooled by it.

When a boiler is properly set, if a fire of a moderate size that burns well does not heat it in a reasonable time, the fault must necessarily lie in the bad management of the doors and registers of the fire-place; for, as the heat required to heat the boiler is *a certain quantity*, which cannot vary, if the boiler is not found to be heated as fast as it ought to be by the quantity of fuel consumed, a part of the heat generated must necessarily go to heat something else; and there is nothing at hand that can take it, except it be the cold air of the atmosphere, which, whenever it is permitted to enter a fire-place in an improper manner or in too large quantities, never fails to rob it of a great deal of heat, which it takes with it up the chimney, as has already been observed.

If the door by which the fuel is introduced into the closed fire-place of a kitchen boiler is not kept constantly closed, it is quite impossible that a well-constructed fire-place can answer. With such neglectful management, *a bad fire-place* is certainly *preferable* to a good one; for, when an enormous quantity of fuel is consumed under a boiler, some part of it must necessarily find its way into it, even if, instead of being set in brick-work, it were suspended over the fire in the open air; but, when a fire-place is made no larger than is necessary in order to heat the boiler in a proper time when the door of the fire-place is kept closed, it is not surprising that the boiler should be much slower in acquiring heat when a stream of cold air is permitted to strike against its bottom and blow all the flame and hot smoke out of its flues into the chimney.

It would be just as unreasonable to object to the fire-places I have recommended, on account of the *trouble of keeping them closed*, as it would be to object to a scheme for warming a dwelling-house merely because it required that the street door should not be left open. The cases are exactly similar; and, if insisting on the attention of servants in the one case is not unreasonable, it cannot be so in the other.

There was a time, no doubt (when the doors of rooms first came in fashion), that the trouble they occasioned to servants was considered as a hardship and severity in exacting attention to the proper management of them as a grievance; but all improvements are progressive, and we may hope that a time will come when it will be considered as careless and slovenly to leave open the door of a closed fire-place. In the mean time, it is my duty to declare, in the *most serious and public manner*, that those who have not influence enough with their servants to secure due attention being paid to this important point, would do wisely not to attempt to introduce the improvements in closed fire-places which I have recommended. And it is not sufficient merely to be attentive to the shutting of the fire-place door. Care must be taken also to manage properly the register of the ash-pit door; otherwise, if it be left too much opened, a great deal too much cold air will find its way into the fire-place between the bars of the grate.

When a closed fire-place is properly constructed, it is hardly to be believed how small a passage is sufficient to admit as much air as is necessary or useful to maintain the combustion of the fuel.

A fault which is often committed in the management of the closed fire-places I have recommended is the *overloading them with fuel.* This mistake has several bad consequences, and among them there is one which would not naturally be expected. It prolongs the kindling of the fire, and very frequently so much so as to prolong the heating of the boiler, notwithstanding the fierceness of the fire when the fuel is all inflamed.

Great care should at all times be taken not to overcharge a fire-place with fuel, but more especially when the fire is first kindled and the fire-place and every thing about it is cold. It should be remembered that a great deal of heat is necessary to warm the fuel itself, and bring it to that degree of heat which it must have in order to its being capable of taking fire; and, as long as there remains any cold fuel in the fire-place to be heated, very little heat will reach the bottom of the boiler.

All the money that is expended in the purchase of wood to kindle coal fires is money well laid out; and it is by no means good economy to be sparing of wood in kindling such fires. In many cases it would, I am convinced, be cheaper to burn wood than coals, even in London, especially in the closed fire-places of small kitchen boilers and stewpans, where a fire is wanted but for a short time. This proposal to burn wood instead of coals or charcoal has already been made more than once; and the more I have considered the subject, the more I am convinced that the former would turn out to be the cheapest fuel.

A great deal of fuel is consumed in this country for boiling water to make tea. I was curious to know how low it would be possible to reduce that expense, and ascertained that point by the following experiments and computations.

I supposed a small family, consisting of two persons, to drink tea twice every day (morning and evening) during one whole year, and that 2 pints of water, at the temperature of 55° (the mean annual temperature of the atmosphere in Great Britain), was heated and made to boil every time tea was made.

I found on inquiry that the most costly fire-wood that is sold in London, — dry beech in billets, — at the highest price it is ever sold at, cost one farthing per lb., avoirdupois weight; that is, at the rate of *twopence* per billet, weighing at an average 8 lbs. By wholesale, these billets are sold in London at *one penny half-penny* each.

I had some of these billets sawed into lengths of about 5 inches, and then split into small pieces (about the size of the end of one's little finger), and bound up with a pack-thread into little small bundles weighing about 4 or 5 ounces each. In the middle of each bundle there were a few smaller splinters and a very small piece of paper, that the bundle might easily be set on fire with a candle or with a common match.

On using the small portable furnace represented in the Fig. 63, and described in Chapter XI. of the tenth Essay,[4] page 310, and the small tin tea-kettles represented in the Fig. 68, in that Essay, I found by an experiment, which was repeated several times, that I could boil 2 pints of water with a bundle of wood weighing 4 ounces.

Hence it appears that the daily consumption of wood in boiling water for tea for two persons would be 8 ounces, or half a pound weight; consequently, for

one year, or 365 days, 182½ lbs. would be required, and
that quantity, at 1 farthing the pound, would cost 182½
farthings $= 45\frac{5}{8}$ pence, or *three shillings* and *ninepence
half-penny* and *half a farthing.*

Were it possible to heat so small a quantity of water
with the consumption of the same proportion of fire-
wood as was found to be sufficient for heating water
in some of the experiments, of which an account is given
in the sixth Essay, the annual expense for fire-wood, for
boiling water for making tea for two persons twice a
day, would amount to no more than 57 lbs. weight,
which, at the London price of this wood, one farthing
in the pound, would cost 57 farthings, or *one shilling
and twopence farthing.*

It is by computations of this sort, founded on the
results of unexceptionable experiments, that we are
enabled to appreciate the vast saving to individuals
and to the public that would result from proper attention being paid to the management of fire and to the
economy of heat.

ON THE

# CONSTRUCTION OF KITCHEN FIRE-PLACES

## AND KITCHEN UTENSILS;

TOGETHER WITH

REMARKS AND OBSERVATIONS RELATING TO THE
VARIOUS PROCESSES OF COOKERY,

AND

PROPOSALS FOR IMPROVING THAT MOST USEFUL ART.

# ADVERTISEMENT.

———————

ALMOST four years have elapsed since this Essay was announced to the public; and although a considerable part of the manuscript was then ready, yet, from a variety of considerations, I have been induced to defer sending it to the press, and even now the first part only of the Essay is laid before the public.

Among the motives which have operated most powerfully to induce me to postpone the publication of this work was a desire to make it as free of faults as possible, and to accommodate it as much as possible to the actual state of opinions and practices in this country.

In proportion as my exertions to promote useful improvements have been favourably received by the public, and my writings have obtained an extensive circulation, my anxiety has been increased to deserve that confidence which is essential to my success. I feel it to be more and more my duty to proceed slowly, and to use every precaution in investigating the subjects I have undertaken to treat, and in explaining what I recommend, in order that others may not be led into errors, either by mistakes in principle or inaccuracy in description.

I have, indeed, of late seen but too many proofs of the necessity of adopting this cautious method of proceeding.

On my return to England from Bavaria last autumn (1798), after an absence of two years, I was not a little gratified to learn that several improvements recommended in my Essays, and particularly the alterations in the construction of chimney fire-places, that were proposed in my fourth Essay, had been adopted in many places, and that they had in general been found to answer very well; but the satisfaction which this information naturally afforded me has since been, I believe I may say, more than counterbalanced by the pain I have experienced on discovering, on a nearer examination, the numerous mistakes that have been committed by those who have undertaken to put my plans in execution; not to mention the unjustifiable use that has in some instances been made of my name in bringing forward for sale inventions which I never recommended, and of which I never can approve without abandoning all the fundamental principles relative to the combustion of fuel, and the management and direction of heat, which, after a long and patient investigation, I have been induced to adopt.

It would be foolish for me to imagine, and ridiculous to pretend, that the plans I have proposed are so perfect as to be incapable of farther improvement. I am far, very far, from being of that opinion, and I can say with truth that I shall at all times rejoice when farther improvements are made in them; but still I may be permitted to add that it would be a great satisfaction to me if those who, from an opinion of their utility or from a desire to give the experiment a fair trial, should be disposed to adopt any of the plans I have recommended, would take the trouble to examine whether the workmen they employ really understand and are disposed to follow the directions I have given; or whether they are not, perhaps, prepossessed with some favourite contrivance and imaginary improvement of their own; or whether there is no danger of their introducing alterations for the purpose of enhancing the price of their work, or of the articles they furnish.

These are dangers of which those who have the smallest acquaintance with mankind must be perfectly sensible; and it would be unwise, and I had almost said unjust, not to attend to them, at least to a certain degree.

All I ask is that a *fair trial* may be given to the plans I propose, when *any* trial is given them; and this request will not, I trust, be thought unreasonable. And as I never presume to recommend to the public any new invention or improvement that I have not previously and repeatedly tried, and found *by experience* to be useful, it would perhaps be thought excusable were I to express a wish that my proposals might not be condemned nor neglected merely in consequence of the failure of contrivances announced as *improvements* of my plans.

The reader will not be surprised at my extreme anxiety to remove those obstacles which appear to me most powerfully to obstruct and retard the general introduction of the improvements I am labouring to introduce; for anxiety for the success of an undertaking naturally flows from a conviction of its importance, and is always connected with that fervent zeal which important undertakings are so eminently calculated to inspire.

———————

To this second edition of the first part of my tenth Essay I beg leave to add a few words respecting the soup establishments that have lately been formed in London and in other places for feeding the poor.

Many persons in this country are of opinion that a great deal of meat is necessary in order to make a good and wholesome soup; but this is far from being the case in fact. Some of the most savoury and most nourishing soups are made without any meat; and in providing food for the poor it is necessary, on many accounts, to be very sparing in the use of it.

When the poor are fed from a public kitchen, care should be taken to supply them with the cheapest kinds of food, and particularly with such as they can afterwards provide for themselves, at their own dwellings, at a small expense; otherwise the temporary relief that is afforded them in times of scarcity, by selling to them rich and expensive meat soups at reduced prices, will operate as a great and permanent evil to themselves and to society.

The most palatable and the most nourishing soups may, with a little care and ingenuity, be composed with very cheap materials, as has been proved of late by a great number of decisive experiments made upon a large scale in different countries. The soup establishments that have been formed at Hamburg, at Geneva, at Lausanne, and other parts of Switzerland, at Marseilles, and lately at Paris, have all succeeded; and at most of these places the kind of soup that was provided for the poor at Munich has been adopted with but little variation. In some cases a small quantity of salt meat has been used, but this has been merely as a seasoning. The basis of these soups has uniformly been barley, potatoes, and peas or beans; and a small quantity of bread has in all cases been added to the soup when it has been served out.

No ingredient is, in my opinion, so indispensably necessary in the soups that are furnished to the poor as *bread*. It should never be omitted, and certainly not in times of scarcity, because there is no way in which bread will go so far as when it is eaten in soups: for every ounce so used, I am confident that four ounces that would otherwise be eaten by the poor at their homes would be saved. And to this we may add that oaten cakes, and other bread of inferior quality, will answer very well in soups, particularly if it be toasted or fried, and broken or cut into small pieces. If the soup be well seasoned, its taste will predominate, and the taste peculiar to the bread will not be perceived.

A great variety of the most agreeable tastes may be given to soups, at a very small expense; and, if bread be mixed with the soup, mastication will be rendered necessary, and the pleasure that is enjoyed in eating a good meal of it will be greatly prolonged and increased.

It is by no means surprising that prejudices should be strong against soups, in those countries where soups and broths are considered as being merely thin wash, without taste or substance, a pint of which might as easily be swallowed down at a breath as so much water; but these prejudices will vanish when the false impressions which gave rise to them are removed.

Soups may, it is true, be made thick and substantial with meat. But, when this is done, they are neither palatable nor wholesome: they appall and load the stomach, weaken the powers of digestion, and instead of affording wholesome nourishment, strength, and refreshment, are the cause of many disorders. They are, moreover, very expensive. But this is not the case with soups made thick and substantial with farinaceous matter, and other vegetable substances, and seasoned and rendered palatable with salt, pepper, onions, and a little salted herrings, hung beef, bacon, or cheese, and eaten with a due proportion of bread.

I am the more anxious to recall the attention of the public to this subject at the present time, as the utility of the public kitchens for feeding the poor, which have lately been formed, and are now forming in various parts of the kingdom, must depend very much on the choice of the ingredients used in preparing food, and the manner of combining them which is adopted by those who have the direction of these interesting establishments. The share I have had in bringing these establishments into use, the opinion I entertain of their importance to society, and the anxiety I must naturally feel for their success, will, I flatter myself, be considered as a sufficient excuse for my solicitude in watching over their progress, and for the liberty I may take in pointing out any mistakes in the management of them that might tend to bring them into disrepute.

# ON THE CONSTRUCTION OF KITCHEN FIRE-PLACES AND KITCHEN UTENSILS.

## INTRODUCTION.

IN contriving machinery for any purpose, it is indispensably necessary to be acquainted with the nature of the mechanical operation to be performed; and though the processes of cookery appear to be so simple and easy to be understood, that any attempt to explain and illustrate them might perhaps be thought not only superfluous, but even frivolous, yet when we examine the matter attentively we shall find their investigation to be of serious importance. I say of *serious* importance; for surely those inquiries which lead to improvements by which the providing of *food* may be facilitated are matters of the highest concern to mankind in every state of society.

The process by which food is most commonly prepared for the table — boiling — is so familiar to every one, and its effects are so uniform, and apparently so simple, that few, I believe, have taken the trouble to inquire *how* or in *what manner* those effects are produced; and whether any and what improvements in that branch of cookery are possible. So little has this matter been an object of inquiry, that few, very few indeed, I believe, among the *millions of persons* who for so many ages have been *daily* employed in this process, have ever given themselves the trouble to bestow one serious thought on the subject.

63

The cook knows, *from experience*, that if his joint of meat be kept a certain time immersed in boiling water it will be *done*, as it is called in the language of the kitchen; but if he be asked *what* is done to it, or *how* or *by what agency* the change it has undergone has been effected, if he understands the question, it is ten to one but he will be embarrassed; if he does not understand it, he will probably answer, without hesitation, that "*the meat is made tender and eatable by being boiled.*" Ask him if the boiling of the water be essential to the success of the process, he will answer, "*Without doubt.*" Push him a little farther, by asking him whether, *were it possible* to keep the water *equally hot* without boiling, the meat would not be cooked *as soon* and *as well* as if the water were made to boil. Here it is probable that he will make the first step towards acquiring knowledge, *by learning to doubt.*

When you have brought him to see the matter in its true light, and to confess that, *in this view of it*, the subject is new to him, you may then venture to tell him (and to prove to him, if you happen to have a thermometer at hand) that water which *just boils* is as hot as it can possibly be made *in an open vessel.* That all the fuel which is used in making it boil with violence is wasted, without adding a single degree to the heat of the water, or expediting or shortening the process of cooking a single instant. That it is by *the heat*, its *intensity* and the *time of its duration*, that the food is cooked, and not by the *boiling* or *ebullition*, or bubbling up of the water, which has *no part whatever* in that operation.

Should any doubts still remain in his mind with respect to the inefficacy and inutility of boiling, in culinary processes, where *the same degree of heat* may be had and be *kept up* without it, let a piece of meat be cooked in a Papin's digester, which, as is well known, is a boiler whose cover (which is fastened down with screws) shuts with so much nicety that no steam can escape out of it. In such a *closed* vessel, boiling (which is nothing else but the escape of steam in bubbles from the hot liquid) is absolutely impossible; yet, if the heat applied to the digester be such as would cause an equal quantity of water in an open vessel to boil, the meat will not only be *done*, but it will be found to be dressed in a shorter time, and to be much tenderer than if it had been boiled in an open boiler. By applying a still greater degree of heat to the digester, the meat may be so much done in a very few minutes as actually to fall to pieces; and even the very bones may be made soft.

Were it a question of mere idle curiosity, whether it be the *boiling* of water, or simply the *degree of heat* which exists in boiling water, by which food is cooked, it would doubtless be folly to throw away time in its investigation; but this is far from being the case, for *boiling* cannot be carried on without a very great expense of fuel; but any boiling-hot liquid (by using proper means for confining the heat) may be kept *boiling-hot* for any length of time almost without any expense of fuel at all.

The waste of fuel in culinary processes, which arises from making liquids boil *unnecessarily*, or when nothing more would be necessary than to keep them *boiling-hot*, is enormous. I have not a doubt but that much more than half the fuel used in all the kitchens, public and private, in the whole world, is wasted precisely in this manner.

But the evil does not stop here.  This unscientific and slovenly manner of cooking renders the process much more laborious and troublesome than otherwise it would be ; and (what by many will be considered of more importance than either the waste of fuel or the increase of labour to the cook) the food is rendered less savoury, and very probably less nourishing and less wholesome.

It is natural to suppose that many of the finer and more volatile parts of food (those which are best calculated to act on the organs of taste) must be carried off with the steam when the boiling is violent; but the fact does not rest on these reasonings.  It is *proved* to a demonstration, not only by the agreeable fragrance of the steam which rises from vessels in which meat is boiled, but also from the strong flavour and superior quality of soups which are prepared by a long process over a very gentle fire.

In many countries, where soups constitute the principal part of the food of the inhabitants, the process of cooking lasts from one meal-time to another, and is performed almost without either trouble or expense. As soon as the soup is served up, the ingredients for the next meal are put into the pot (which is never suffered to cool, and does not require scouring); and this pot, — which is of cast iron or of earthen-ware, — being well closed with its thick wooden cover, is placed *by the side of the fire*, where its contents are kept simmering for many hours, but are seldom made to boil, and never but in the gentlest manner possible.

Were the pot placed in a closed fire-place (which might easily be constructed, even with the rudest materials, with a few bricks or stone, or even with sods, like a camp-kitchen), no arrangement for cooking could well be imagined more economical or more convenient.

Soups prepared in this way are uncommonly savoury; and I am convinced that the true reason why nourishing soups and broths are not more in use among the common people in Great Britain and Ireland is because they do not know how good they really are, nor how to prepare them; in short, because they are not acquainted with them.

But to return from this digression. It is most certain not only that meat and vegetables of all kinds may be cooked in water which is kept *boiling-hot* without actually boiling, but also that they may even be cooked with a degree of heat *below* the boiling point.

It is well known that the heat of boiling water is not the same in all situations, — that it depends on the pressure of the atmosphere, and consequently is considerably greater at the level of the surface of the sea than inland countries, and on the tops of high mountains; but I never heard that any difficulty was found to attend the process of dressing food by boiling, even in the highest situations. Water boils at London (and at all other places on the same level) at the temperature of 212 degrees of Fahrenheit's thermometer; but it would be absolutely impossible to communicate that degree of heat to water in an open boiler in Bavaria. The boiling-point at Munich, under the mean pressure of the atmosphere at that place, is about $209\frac{1}{2}$ degrees of Fahrenheit's thermometer; yet nobody, I believe, ever perceived that boiled meat was *less thoroughly done* at Munich than at London. But if meat may, without the least difficulty, be cooked with the heat of $209\frac{1}{2}$ degrees of Fahrenheit at Munich, why should it not be possible to cook it with the same degree of heat in London? If this can be done (which I think can hardly admit of a doubt), then it is evident that the process of cookery, which is called *boiling*, may be performed in water which is not boiling-hot.

I well know, from my own experience, how difficult it is to persuade cooks of this truth; but it is so important, that no pains should be spared in endeavouring to remove their prejudices and enlighten their understandings. This may be done most effectually in the case before us by a method I have several times put in practice with complete success. It is as follows: Take two equal boilers, containing equal quantities of *boiling-hot water*, and put into them two equal pieces of meat taken from the same carcass, — two legs of mutton, for instance, — and boil them during the same time. Under one of the boilers make a *small fire*, just barely sufficient to keep the water *boiling-hot*, or rather just *beginning to boil;* under the other make *as vehement a fire as possible*, and keep the water boiling the whole time with the utmost violence.

The meat in the boiler in which the water has been kept *only just boiling-hot* will be found to be quite as well done as that in the other,* under which so much fuel has been wasted in making the water boil violently to no useful purpose. It will even be more done; for, as a great deal of water will be boiled away (evaporated) during the process in the boiler under which a great fire is kept up, this boiler must often be filled up; and, if the water with which it is from time to time replenished be cold, this will of course retard the process of cooking the meat.

---

* It will even be found to be much better cooked; that is to say, tenderer, more juicy, and much higher flavoured.

To form a just idea of the enormous waste of fuel that arises from making water boil, and *evaporate* unnecessarily in culinary processes, we have only to consider how much heat is expended in the formation of steam. Now it has been proved by the most decisive and unexceptionable experiments that have ever been made by experimental philosophers that, if it were possible that the heat which actually combines with water in forming steam (and which gives it wings to fly up into the atmosphere) could exist in the water without changing it from a dense liquid to a rare elastic vapour, this water would be heated by it to the temperature of red-hot iron.

From the same *data* it is easy to show by computation that, if any given quantity of ice-cold water can be made to boil with the heat generated in the combustion of a certain quantity of any given kind of fuel, it will require more than *five times* that quantity of fuel to reduce that same quantity of water — already boiling-hot — to steam.

Hence it appears that, in the formation of steam, there is a great and unavoidable *expense* of heat; but it does not seem probable that heat is *expended* or *combined* in any of those processes by which food is prepared for the table, except it be, perhaps, in baking; and as heat is *immortal*, — that is to say, as it never dies or ceases to exist, and as its dispersion may be prevented, or at least greatly *retarded*, by various simple contrivances, — it is not surprising, when we consider the matter attentively, that most of those processes (in which nothing more seems to be necessary than that the food to be cooked should be exposed a certain time in a medium at a certain temperature) should be capable of being performed with *a very small expense of fuel.*

The quantity of heat, or rather the quantity of fuel, by which any given culinary process may be performed, may be determined with much certainty and precision from the results of experiments which have already been made.

Suppose, for instance, it were required to compute the quantity of dry pine-wood (what, in England, is called deal) used as fuel, and burned in a closed fire-place, constructed on the most approved principles, to boil 100 lbs. of beef. And, first, we will suppose this beef to be in such large pieces that 3 hours of boiling, after it has been made boiling-hot, are necessary to make it sufficiently tender to be fit for the table ; and we will suppose, farther, that 3 lbs. of water are necessary to each pound of beef, and that both the water and the beef are at the temperature of 55° of Fahrenheit's thermometer (the mean temperature of the atmosphere in England) at the beginning of the experiment.

The first thing to be ascertained is how much fuel would be required to heat the water and the beef *boiling-hot ;* and then to see how much more would be required to *keep them boiling-hot* three hours.

And, first, for *heating the water.* It has been shown by one of my experiments (No. 20; Vol. II, p. 387) that $20\frac{1}{10}$ lbs. of water may be heated 180 degrees of Fahrenheit's thermometer with the heat generated in the combustion of 1 lb. of dry pine-wood.

But it is required to heat the water in question only 157 degrees ; for its temperature being that of 55°, and the boiling-point 212°, it is $212° - 55° = 157°$; and if 1 lb. of the fuel be sufficient for heating $20\frac{1}{10}$ lbs. of water 180 degrees, it must be sufficient for heating 23 lbs. of water 157 degrees, for 157° is to 180° as $20\frac{1}{10}$ lbs. to 23 lbs.

But if 23 lbs. of water, at the temperature of 55°, require 1 lb. of dry pine-wood, as fuel, to make it boil, then 300 lbs. of water (the quantity required in the process in question) would require $12\frac{6}{10}$ lbs. of the wood to heat it boiling-hot.

To this quantity of fuel must be added that which would be required to heat the meat (100 lbs. weight) boiling-hot. Now it has been found by actual experiment by the late ingenious Doctor Crawford (see his Treatise on Animal Heat, second edition, page 490) that the flesh of an ox requires less heat to heat it than water, in the proportion of 74 to 100; consequently the quantity of beef in question (100 lbs.) might be made boiling-hot with precisely the same quantity of fuel as would be required to heat 74 lbs. of water at the same temperature to the boiling-point. And this quantity in the case in question would amount to $3\frac{1}{4}$ lbs., as will be found on making the computation.

This quantity ($3\frac{1}{4}$ lbs.) added to that before found, which would be required to heat the water alone (= 23 lbs.), gives $26\frac{1}{4}$ lbs. of dry pine-wood for the quantity required to heat 300 lbs. of water and 100 lbs. beef (both at the temperature of 55°) boiling-hot.

To estimate the quantity of fuel which would be necessary to keep this water and beef boiling-hot 3 hours, we may have recourse to the results of my experiments. In the Experiment No. 25 (Vol. II, p. 389), 508 lbs. of boiling-hot water were kept actually boiling — not merely kept boiling-hot — 3 hours with the heat generated in the combustion of $4\frac{1}{2}$ lbs. of dry pine-wood:

this gives 338⅔ lbs. of boiling-hot water kept boiling 1 hour with 1 lb. of the fuel; and computing from these data, and supposing, farther, that a pound of beef requires as much heat to keep it boiling-hot any given time as a pound of water, it appears that 3½ lbs. of pine-wood, used as fuel, would be sufficient to keep the 300 lbs. of water, with the 100 lbs. of beef in it, boiling 3 hours. This quantity of fuel ($= 3\frac{1}{2}$ lbs.), added to that required to heat the water and the meat boiling-hot ($= 26\frac{1}{2}$ lbs.), gives 29¾ lbs. of pine-wood for the quantity of fuel required to cook 100 lbs. of boiled beef.

This quantity of fuel, which is just about equal in effect to 16 lbs., or ¾ of a peck of pit-coal, will doubtless be thought a small allowance for boiling 100 lbs. of beef; but it is in fact much more than would be necessary *merely for that purpose*, could all the heat generated in the combustion of the fuel be applied *immediately* to the cooking of the meat, and *to that purpose alone*. Much the greatest part of that which is generated is expended in heating the water in which the meat is boiled, and as it remains in the water after the process is ended it must be considered as lost.

This loss may, however, be prevented in a great measure; and, when that is done, the expense of fuel in boiling meat will be reduced almost to nothing. We have just seen that 100 lbs. of meat, at the mean temperature of the atmosphere in England (55°), may be made boiling-hot with the heat generated in the combustion of 3¼ lbs. of pine-wood; and there is no doubt but, with the use of proper means for confining the heat, this meat might be kept boiling-hot 3 hours, and consequently be thoroughly done, with the addition of $\frac{3}{4}$ of a pound of the fuel, making in all 4 lbs. of pine-wood, equal in effect to about $2\frac{1}{4}$ lbs. of pit-coal; which, according to this estimate, is all the fuel that would be *absolutely necessary* for cooking 100 lbs. of beef.

This quantity of fuel would cost in London less than *one farthing and a half*, when the chaldron of coals weighing 28 cwt. is sold at 40 shillings. This, however, is the *extreme* or *utmost limit* of the economy of fuel, beyond which it is absolutely impossible to go. It is even impossible, in practice, to arrive at this limit, for the containing vessel must be heated, and kept hot, as well as the meat; but very considerable advances may be made towards it, as I shall show hereafter.

If we suppose the meat to be boiled in the usual manner, and that 300 lbs. of cold water are heated expressly for that purpose, in that case the fuel required, amounting to 16 lbs. of coal, would cost in London (the chaldron reckoned as above) just *2 pence* $1\frac{3}{4}$ *farthings*. But all this expense ought not to be placed to the account of the cooking of the meat. By adding a few pounds of barley meal, some greens, roots, and seasoning to the water, it may be changed into a good and wholesome soup, at the same time that the meat is boiled; and the expense for fuel (2 pence $1\frac{3}{4}$ farthings) may be divided between the meat boiled (100 lbs.) and 300 lbs., or $37\frac{1}{2}$ gallons, of soup.

I am aware of the danger to which I expose myself by entertaining the public with accounts of facts, and of deductions from them, which are certainly much too new and extraordinary to be credited but on the strongest proofs, while many of the arguments and computations I offer in their support — however con-clusive they may, and certainly *must*, appear to natural philosophers and mathematicians — are such as the generality of readers will be tempted to pass over without examination; but, deeply impressed with the importance of the object I have in view, I am determined to pursue it at all hazards.

My principal design in publishing these computations is to *awaken the curiosity of my readers*, and fix their attention on a subject which, however low and vulgar it has hitherto generally been thought to be, is in fact highly interesting, and deserving of the most serious consideration. I wish they may serve to inspire cooks with a just idea of the importance of their art, and of the intimate connection there is between the various processes in which they are daily concerned, and many of the most beautiful discoveries that have been made by experimental philosophers in the present age.

The advantage that would result from an application of the late brilliant discoveries in philosophical chemistry, and other branches of natural philosophy and mechanics, to the improvement of the art of cookery, are so evident and so very important that I cannot help flattering myself that we shall soon see some enlightened and liberal-minded person of the profession take up the matter in earnest, and give it a thoroughly *scientific* investigation.

In what art or science could improvements be made that would more powerfully contribute to increase the comforts and enjoyments of mankind?

And it must not be imagined that the saving of fuel is the only or even the most important advantage that would result from these inquiries: others of still greater magnitude, respecting the *manner* of preparing food for the table, would probably be derived from them.

The heat of boiling water, continued for a shorter or a longer time, having been found by experience to be sufficient for cooking all those kinds of animal and vegetable substances that are commonly used as food; and *that degree* of heat being easily procured, and easily kept up, in all places and in all seasons; and as all the utensils used in cookery are contrived for that kind of heat, few experiments have been made to determine the effects of using *other degrees of heat*, and *other mediums* for conveying it to the substance to be acted upon in culinary processes. The effects of different degrees of heat in the same body are, however, sometimes very striking; and the taste of the same kind of food is often so much altered by a trifling difference in the manner of cooking it, that it would no longer be taken for the same thing. What a surprising difference, for instance, does the manner of performing that most simple of all culinary processes, *boiling in water*, make on potatoes! Those who have never tasted potatoes *boiled in Ireland*, or cooked according to the Irish method, can have no idea what delicious food these roots afford when they are properly prepared. But it is not merely the *taste* of food that depends on the manner of cooking it: its nutritiousness also, and its wholesomeness, — qualities still more essential if possible than taste, — are, no doubt, very nearly connected with it.

Many kinds of food are known to be most delicate and savoury when cooked in a degree of heat considerably below that of boiling water; and it is more than probable that there are others which would be improved by being exposed in a heat greater than that of boiling water.

In the seaport towns of the New England States in North America, it has been a custom, time immemorial, among people of fashion, to dine one day in the week (Saturday) on *salt-fish;* and a long habit of preparing the same dish has, as might have been expected, led to very considerable improvements in the art of cooking it.    I have often heard foreigners, who have assisted at these dinners, declare that they never tasted salt-fish dressed in such perfection; and I well remember that the secret of cooking it is to keep it a great many hours in water that is *just scalding-hot*, but which is never made actually to boil.

I had long suspected that it could hardly be possible that *precisely* the temperature of 212 degrees of Fahrenheit's thermometer (that of boiling water) should be that which is best adapted for cooking *all sorts of food;* but it was the unexpected result of an experiment that I made with another view which made me particularly attentive to this subject. Desirous of finding out whether it would be possible to roast meat in a machine I had contrived for drying potatoes, and fitted up in the kitchen of the House of Industry at Munich, I put a shoulder of mutton into it, and after attending to the experiment three hours, and finding it showed no signs of being done, I concluded that the heat was not sufficiently intense; and, despairing of success, I went home rather out of humour at my ill success, and abandoned my shoulder of mutton to the cook-maids.

It being late in the evening, and the cook-maids thinking, perhaps, that the meat would be as safe in the drying-machine as anywhere else, left it there all night. When they came in the morning to take it away, intending to cook it for their dinner, they were much surprised to find it *already cooked*, and not merely eatable, but perfectly done, and most singularly well-tasted.    This appeared to them the more miraculous, as the fire under the machine was gone quite out before they left the kitchen in the evening to go to bed, and as they had locked up the kitchen when they left it and taken away the key.

This wonderful shoulder of mutton was immediately brought to me in triumph, and though I was at no great loss to account for what had happened, yet it certainly was quite unexpected; and when I tasted the meat I was very much surprised indeed to find it very different, both in taste and flavour, from any I had ever tasted. It was perfectly tender; but, though it was so much done, it did not appear to be in the least sodden or insipid, — on the contrary, it was uncommonly savoury and high flavoured. It was neither boiled nor roasted nor baked. Its taste seemed to indicate the manner in which it had been prepared; that the gentle heat, to which it had for so long a time been exposed, had by degrees loosened the cohesion of its fibres, and concocted its juices, without driving off their fine and more volatile parts, and without washing away or burning and rendering rancid and empyreumatic its oils.

Those who are most likely to give their attention to this little history will perceive what a wide field it opens for speculation and curious experiment. The circumstances I have related, however trifling and uninteresting they may appear to many, struck me very forcibly, and recalled to my mind several things of a similar nature which had almost escaped my memory. They recalled to my recollection the manner just described in which salt-fish is cooked in America; and also the manner in which *samp* is prepared in the same country. (See my Essay on Food.)[5] This substance, which is exceedingly palatable and nourishing food when properly cooked, *is not eatable* when simply boiled. How many cheap articles may there be of which the most delicate and wholesome food might be prepared, were the art and the *science* of cooking them better understood. But I beg my reader's pardon for detaining him so long with speculations which he may perhaps consider as foreign to the subject I promised to treat in this Essay. To proceed, therefore, to those investigations which are more immediately connected with the construction of kitchen fire-places.

# PART I.

---

## CHAPTER I.

*Of the Imperfections of the Kitchen Fire-places now in common Use. — Objects particularly to be had in View in Attempts to improve them. — Of the Distribution of the various Parts of the Machinery of a Kitchen. — Of the Method to be observed in forming the Plan of a Kitchen that is to be fitted up, and in laying out the Work.*

A S the principal object of this publication is to convey such plain and simple directions for constructing kitchen fire-places and kitchen utensils as may easily be understood, even by those who are not versed in philosophical inquiries, and who have not had leisure to examine scientifically the principles on which the proposed improvements are founded, I shall endeavour, in treating the subject, to make use of the plainest language, and to avoid as much as possible all abstruse and difficult investigation.

It will be proper to begin by taking a cursory view of kitchen fire-places, as they are now commonly constructed, and to point out their defects, and show what the objects are which ought principally to be had in view in attempts to improve them.

79

*Of the Imperfections of the Kitchen Fire-places now in common Use.*

The great fault in the construction and arrangement of the kitchens of private families now in common use in most countries, and particularly in Great Britain and Ireland (a fault from which all their other imperfections arise), is that they are not *closed*. The fuel is burned in a long open grate called a *kitchen range*, over which the pots and kettles are freely suspended, or placed on stands; or fires are made with charcoal in square holes, called *stoves* in a solid mass of brick-work, and connected with no flue to carry off the smoke, over which holes stewpans or saucepans are placed on tripods, or on bars of iron, exposed on every side to the cold air of the atmosphere.

The loss of heat and waste of fuel in these kitchens is altogether incredible; but there are other evils attending them, which are, perhaps, still more important. All the various processes in which fire is used in preparing food for the table are extremely unpleasant and troublesome in these kitchens, not only on account of the excessive heat to which those are exposed who are employed in them, but also and more especially on account of the *noxious exhalations* from the burning charcoal, and the *currents of cold air* in the kitchen, which are occasioned by the strong draught up the chimney.

It is sufficient to have once been in a kitchen when dinner was preparing for a large company, or even merely to have met the cook coming sweltering out of it, to be convinced that the business of cooking, as it is now performed, is both disagreeable and unwholesome;

and it appears to me that it would be no small addition to the enjoyments of those who are fond of the pleasures of the table to know that they were procured with less trouble and with less injury to the health of those who are employed in preparing them.

Another inconvenience attending open chimney fire-places, as they are now constructed, is the great difficulty of preventing their smoking. In order that there may be room for all the pots and kettles which are placed over the fire, the grate, or *kitchen range*, as it is called, must be very long; and in order that the cook may be able to approach these pots, etc., the mantel of the chimney is made very high: consequently the throat of the chimney is not only enormously large, but it is situated very high above the burning fuel, both of which circumstances tend very much to make a chimney smoke, as I have shown in my Essay on Open Chimney Fire-places;[6] and there does not appear to be any effectual remedy for the evil, without altering entirely the construction of such fire-places.

### *Of the Objects particularly to be had in View in Attempts to improve Kitchen Fire-places.*

The objects which ought principally to be attended to in the arrangement of a kitchen are the following: —

1*st*, Each boiler, kettle, and stewpan should have its separate closed fire-place.

2*dly*, Each fire-place should have its grate, on which the fuel must be placed, and its separate ash-pit, which must be closed by a door well fitted to its frame, and furnished with a register for regulating the quantity of air admitted into the fire-place through the grate. It should also have its separate canal for carrying off the smoke into the chimney, which canal should be furnished with a damper. By means of this damper and of the ash-pit door register, the rapidity of the combustion of the fuel in the fire-place, and consequently the rapidity of the generation of the heat, may be regulated at pleasure. The economy of fuel will depend principally on the proper management of these two registers.

3*dly*, In the fire-places for all boilers and stewpans which are more than 8 or 10 inches in diameter, or which are too large to be easily removed with their contents *with the strength of one hand*, a horizontal opening just above the level of the grate must be made for introducing the fuel into the fire-place, which opening must be nicely closed by a fit stopper or by a double door. In the fire-places which are constructed for smaller stewpans this opening may be omitted, and the fuel may be introduced through the same opening into which the stewpan is fitted, by removing the stewpan occasionally for a moment for that purpose.

4*thly*, All portable boilers and stewpans, and especially such as must often be removed from their fire-places, should be circular, and they should be suspended in their fire-places by their circular rims; but the best form for all fixed boilers, and especially such as are very large, is that of an oblong square, and all boilers, great and small, should rather be broad and shallow than narrow and deep.

A circular form is best for portable boilers, on account of the facility of fitting them to their fire-places; and an oblong square form is best for large fixed boilers, on account of the facility of constructing and repairing the straight horizontal flues under them and round them, in which the flame and smoke by which they are heated are made to circulate.

When large boilers are shallow, and when their bottoms are supported on the tops of narrow flues, the pressure or weight of their contents being supported by the walls of the flues, the metal of which the boiler is constructed may be very thin, which will not only diminish very much the first cost of the boiler, but will also greatly contribute to its durability; for the thinner the bottom of a boiler is, the less it is fatigued and injured by the action of the fire, and the longer, of course, it will last; which is a curious fact, that has hitherto been too little known, or not enough attended to, in the construction of large boilers.

5*thly*, All boilers, great and small, should be furnished with covers, which covers should be constructed in such a manner and of such materials as to render them well adapted for confining heat. Those who have never examined the matter with attention would be astonished on making the experiment to find how much heat is carried off by the cold air of the atmosphere from the surface of hot liquids, when they are exposed naked to it, in boilers without covers. But in culinary processes it is not merely the loss of heat which is to be considered: a great proportion of the finer and more rich and savoury particles of the food are also carried off at the same time, and lost, which renders it an object of serious importance to apply an effectual remedy to this evil.

As heat makes its way through wood with great difficulty, and very slowly, there would perhaps be no substance better adapted for constructing covers for boilers than it, were it not for the perpetual changes in its form and dimensions which are occasioned by alternate changes of dryness and moisture; but these alterations are so considerable, and their effects so difficult to be counteracted, especially when the form of the cover is circular, that, for portable boilers and for stewpans and saucepans, I should prefer covers made of thin sheets of tinned iron, or of tin, as it is commonly called.   These covers (which must always be made double) have already been particularly described in my sixth Essay.[3]

Though boilers and stewpans should never be used naked over an open fire, or otherwise than in closed fire-places, yet it is not necessary in fitting up a kitchen to build as many separate fire-places as it may be proper to have boilers, stewpans, and saucepans; for the same fire-place may be made to serve occasionally for several boilers or stewpans.   Those, however, that are used in the same closed fire-place must be all of the same diameter; and, in order that their capacities may be different, they may be made of different depths.

As, in the hurry of business in the kitchen, one stewpan or boiler might easily be taken for another, were their diameters to vary by only a small difference, and were they not distinguished by marks or numbers, — to prevent these mistakes, their diameters, expressed in inches, should be marked on some conspicuous part, — on their handles for instance, or on their brims, and also on their covers; and their fire-places should be marked with the same number.

To guard still more effectually against all mistakes respecting the sizes of these utensils, and the fire-places to which they belong, the difference of the diameters of two boilers or stewpans should never be less than one whole inch.   In several private kitchens that have been constructed on my principles, their diameters have been made to vary by two inches, — that is to say, they have been made of 6, 8, 10, 12, and 14 inches in diameter; and, in order that those of the same diameter might be of different capacities, they were made of three different depths, namely, $\frac{1}{3}$, $\frac{1}{2}$, and $\frac{2}{3}$ their diameter in depth. Not only the numbers which show their diameters, but the fractions also which express their depths, are marked on their handles, or on their brims.

The size of a private kitchen, or the number and size of its separate closed fire-places, and of its boilers and stewpans, must be regulated by the size of the family, or rather by the style of living; for, where sumptuous entertainments are occasionally provided for large companies, the kitchen must be spacious and its arrangement complete, however small the family may be, or however moderate the expenses of their table may be in their ordinary course of living in private.

Yet when kitchens are fitted up on the principles I am desirous of recommending, neither the size of the kitchen, nor the number or dimensions of its utensils, will occasion any addition to the table expenses of the family in their ordinary course of living when they have no company, which is an important advantage that these kitchens have over those on the common construction.

In large kitchens with open fire-places, the kitchen range being wide and very roomy, an enormous quantity of fuel is swallowed up by it, even when only a very small quantity of food is provided; but this unnecessary waste is completely prevented by cooking in boilers and stewpans properly fitted into separate closed fire-places.

More fuel is frequently consumed in a kitchen range to boil a tea-kettle than, with proper management, would be sufficient to cook a dinner for fifty men.

### *Of the Distribution of the various Parts of the Machinery of a Kitchen.*

Though the internal construction of the fire-places, and the means employed for confining and directing the heat generated in the combustion of the fuel (subjects which have been thoroughly investigated in my sixth Essay), are matters of the first concern in the fitting up of a kitchen, yet these are not all that require attention. The distribution of the various parts of the machinery is a matter of considerable importance, for a good arrangement of the different instruments and utensils — of the boilers, ovens, roasters, etc. — will tend very much to facilitate the business of cooking, and consequently *to put the cook in good humour*, which is certainly a matter of serious importance.

Cooks in general are averse to all new inventions, and this is not surprising, and ought by no means to be imputed to them as a fault. Accustomed *to work with their own tools*, they naturally feel awkward and embarrassed when others are put into their hands; and to this we may add that there is always a degree of humiliation felt by those who, after having been accustomed to consider themselves, and to be considered by others, as masters of their profession, are required to learn any thing new, or to do any thing in any other manner than that in which they have always been accustomed to do it, and in the performance of which they have always acquired praise. It will not, however, be difficult to convince those of the profession who are possessed of a good understanding, and are above low and vulgar prejudices, that the alterations proposed will most certainly meet with their approbation *when they become better acquainted with them.*

The distribution of the parts of a kitchen must always depend so much on local circumstances that general rules can hardly be given respecting it: the principles, however, on which this distribution ought in all cases to be made — viz., convenience to the cook, cleanliness, and symmetry — are simple, and easy to be understood; and, in the application of them, the architect will have a good opportunity of displaying his ingenuity and showing his taste.

Should he condescend to consult the cook in making these arrangements, he will do wisely, on more accounts than one.

Though the smoke from the fire-places of the boilers may be conveyed almost to any distance in horizontal canals, yet it will in most cases be advisable to place the boilers near the chimney; and it will in general, though not always, be best to place them all in one range, or rather in one mass of brick-work.

*Of the Method of forming a Plan of a Kitchen that is to be fitted up, and of laying out the Work.*

Before the plan of a kitchen which it is intended to fit up is made, an exact plan must be procured of the room in which it is to be constructed, in which plan all the doors and windows must be distinctly marked, and also the fire-place, if there be one in the room, and the chimney. The number and the dimensions must likewise be known of all the boilers and saucepans which are to be fitted up in the brick-work.

The readiest way of proceeding in making a plan or drawing of the machinery of a kitchen is to form it on the plan of the room; and in doing this the work will be much facilitated by the following very simple contrivance.

Cut out of thick pasteboard detached pieces to rep resent the boilers, saucepans, roasters, ovens, etc., which are to be fitted up in the brick-work, and placing these in different ways on the plan of the room, see in what manner they can best be disposed or arranged. As these models (which must be drawn to the same scale as that used in drawing the plan of the room) may be moved about at pleasure, and placed in an infinite variety of different positions in regard to each other, and to the different parts of the room; the effect of any proposed arrangement may be tried in a few moments, in a very satisfactory manner, without expense, and almost without any trouble.

To facilitate still more these preliminary trials with these models of the boilers, etc., several slips of pasteboard, equal in width to the distance at which one boiler ought to be placed from the other in the brick-work, measured on the scale of the plan, should be provided and used in placing the models of the boilers at proper distances from each other. This distance in fitting up or setting kitchen boilers and saucepans I have commonly taken at the width of a brick, or $4\frac{1}{2}$ inches; and I have allowed the same space ($4\frac{1}{2}$ inches) for the distance of the side of the boiler from the outside or front of the mass of the brick-work in which it is set. When this point is settled (that respecting the distance which should be left between the boilers), the arranging of the pasteboard models of the boilers on the plan will be perfectly easy.

As soon as the distribution of the various boilers,

etc., is finally settled, a ground plan of the whole of the machinery should be traced on the plan of the room; and a sufficient number of sections and elevations should be drawn to show the situations, forms, and dimensions of the fire-places, and of all the other parts of the apparatus.

When this is done, and when the boilers and the materials for building are provided, and every thing else that can be wanted in fitting up the kitchen is in readiness, the architect or amateur may proceed to the laying out of the work.

As this will not be found to be difficult, and as it is really a most amusing occupation, I cannot help recommending it very earnestly to gentlemen, and even to ladies, to superintend and direct these works.

I don't know what opinion others may entertain of these amusements, but with regard to myself I own that I know of nothing more interesting than the planning and executing of machinery, by which the powers of Nature are made subservient to my views, by which the very elements are bound as it were in chains, and made to obey my despotic commands; and not my commands alone, but those of all the human race, to whose necessities and comforts they are made the faithful and obedient ministers.

The first thing to be done in laying out the work when a kitchen is to be fitted up is to draw with red or white chalk, or with a coal, a ground plan of the brick-work, of the full size, on the floor or pavement of the room. When the kitchen is neither paved nor floored, this drawing must, of course, be made on the ground. In this drawing, the ash-pits and the passages leading to them must be marked; and, when the ash-pit is to be sunk into the ground, that is the first thing that must be executed.

As soon as this ground plan is sketched out, the ash-pit doors should all be placed, and the foundations of the brick-work laid.

To assist the bricklayer, and prevent his making mistakes, several sections of the brick-work of the full size, and particularly sections of all the boilers, represented as fixed in their fire-places, should be drawn on wide boards, or on very large sheets of paper, or they may be drawn with charcoal or red chalk on the sides of the room. These sections of the full size, where the bricklayer can readily take measure of the various parts of the work to be performed, will be found very useful.

Before I proceed to give a more particular and minute description of the various kitchen utensils and other machinery which will be recommended, I shall lay before my reader an account, illustrated by drawings, of several complete kitchens that have already been constructed under my direction. I have been induced to adopt this method in treating my subject, from an opinion that the directions which still remain to be given respecting the construction of kitchen fire-places and of kitchen utensils will more easily be understood when a general idea shall have been formed of some of those kitchens which have already been constructed on the principles recommended.

# CHAPTER II.

*Detailed Accounts, illustrated by correct Plans, of various Kitchens, public and private, that have already been constructed on the Author's Principles, and under his immediate Direction.*

ONE of the most complete kitchens I have ever yet caused to be constructed is, in my opinion, that belonging to Baron de Lerchenfeld at Munich, and although its general form and the distribution of the machinery are very different from any thing that has been seen in this country, — so different that I should, perhaps, doubt whether it would be prudent at the first outset to recommend their adoption and exact imitation, — yet as this kitchen has been found to answer remarkably well, — even to the entire satisfaction of the cook, who began, however, by entering his formal protest against it, — I have thought it right to lay the following description of it before my readers. Those who are alarmed at the novelty of its appearance will be so good as to recollect that much may be done, as will hereafter be shown, by way of accommodating the plan to the idea of those to whom it is too new not to appear extraordinary and uncouth.

*Description of a Kitchen in the House of Baron de Lerchenfeld at Munich.*

PLATE VII.

Fig. 1. This plate shows a perspective view of the kitchen fire-place seen nearly in front. The mass of

PLATE VII

Fig. I

Perspective View of the Kitchen of Baron de Lerchenfeld at Munich.

brick-work in which the boilers and saucepans are set projects out into the room, and the smoke is carried off by flues that are concealed in this mass of brick-work, and in the thick walls of an open chimney fire-place which, standing on it, on the farther side of it, where it joins to the side of the room, is built up perpendicularly to the ceiling of the room. At the height of about 12 or 15 inches above the level of the mantel of this open chimney fire-place, the separate canals for the smoke concealed in its walls end in the larger canal of this fire-place, which last-mentioned larger canal, sloping backwards, ends in a neighbouring chimney which carries off the smoke through the roof of the house into the atmosphere.

A horizontal section of this open chimney fire-place, at the level of the upper surface of the mass of brick-work on which it stands, may be seen Plate IX., Fig. 5. In this section the vertical canals are distinctly marked, which carry off the smoke from the boilers into the chimney, as also the stoppers which are occasionally taken away to remove the soot, when these canals are cleaned. These stoppers, which are made of earthenware burnt like a brick or tile, are 8 inches long, 6 inches wide, and 3 inches thick, and on their outsides they have two deep grooves that form a kind of handle for taking hold of them. When they are fixed in their places, their joinings with the door-way into which they are fitted are made tight by filling up the crevices with moist clay. The canals are cleaned by means of a strong cylindrical brush, made of hogs' bristles fixed to a long flexible handle of twisted iron wire.

The open chimney fire-place was constructed in order that an open fire might be made on its hearth (which, as appears by the plan, is on a level with or is a continuation of the top or upper surface of the mass of brick-work in which the boilers are set), should any such fire be wanted; but the fact is that, although this kitchen has been in daily use more than five years, it has not yet been found necessary to light a fire in this place. When any thing is to be fried or broiled, the cook finds it very convenient to perform these processes of cookery over the two large stoves that are placed in the front of this open fire-place, as the disagreeable vapour that rises from the frying-pan or from the grid-iron goes off immediately by the open chimney; and these stoves serve likewise occasionally for warming heaters for ironing, and also for burning wood to obtain live coals for warming beds, or for keeping up a small fire for boiling a tea-kettle, or for warming any thing that is wanted in the family. When this fire is not wanted, the register in the ash-pit door is nearly closed, and the top of the stove is covered with a fit cover of earthen-ware, by which means the fire is kept alive for a great length of time, almost without any consumption of fuel; and may at any time be revived and made to burn briskly in less than half a minute, merely by admitting a larger current of fresh air.

The convenience in a family of being able to have a brisk fire in the kitchen in a moment, when wanted, and to check the combustion in an instant, without extinguishing the fire, and without even cooling the fire-place, when the fire is no longer wanted, can hardly be conceived by those who have not been used to any other methods of making and keeping up kitchen fires than those commonly used in the kitchens in Great Britain.

It will certainly be confessed that neither science nor art has done much either for saving labour or for saving expense, either for convenience, comfort, cleanliness, or economy in the invention and management of a *kitchen range*.

Before I proceed to explain more minutely the different parts of this kitchen, it may be useful to give a general idea of the whole of it, taken together.

### PLATE VIII.

Fig. 2. This figure shows a front view, or, more strictly speaking, an elevation of this kitchen. In this plan the ash-pit doors with their registers are distinctly seen; and also the ends of the earthen stoppers which close the openings into the fire-places* of four of the principal boilers. The covers of the principal boilers,† as also of several of the stewpans, are seen above the level of the upper surface of the mass of brick-work.

The height of this mass of brick-work, *a b*, measured from the floor or pavement of the kitchen, is just 3 feet.

Fig. 3. This figure shows a horizontal section of the mass of brick-work in which the boilers, etc., are set, taken at the level of the horizontal flues, that carry off the smoke from the boilers, stewpans, and saucepans, into the vertical canals which convey it into the chimney.

The smoke from three of the principal boilers, situated on the left hand, is carried by separate canals to a circular cavity, over which a large shallow boiler is placed, in which water is heated (by this smoke) for the use of the kitchen, and more especially for washing the plates and

---

\* For a particular account of these stoppers, see Vol. II, pp. 332 and 465, Plate I., Figs. 6, 7, and 8.

† For an account of these covers, see Vol. II, pp. 321 and 465, Plate I., Figs. 1 and 2.

PLATE VIII

_Fig. 2._

_Plans of Baron de Lerchenfeld's Kitchen?_

_Fig. 3._

dishes. This boiler is distinctly seen with its wooden
cover (consisting of three pieces of deal united by two
pairs of hinges) in the Fig. 5, Plate IX.

The five fire-places on the left-hand side of the mass of
brick-work are represented without their circular grates,
and the eight fire-places that are situated on the right
hand are shown with their circular grates in their places.*

The fire-places of the four largest boilers, which are
situated in front of the brick-work, have doors or openings, closed with stoppers, for introducing fuel into these
fire-places, and three of these openings are represented
in the plan as being closed by their stoppers; while the
fourth (that situated on the right hand) is shown open,
or without its stopper.

As all the rest of the fire-places (or stoves, as they
would be called in this country) are without any lateral
opening for introducing the fuel, when any fuel is to be
introduced into one of these fire-places, the stewpan or
saucepan must be removed for a moment for that
purpose.

It will be observed that several of the horizontal canals
that carry off the smoke from the boilers are divided into
two branches, which unite at a little distance from their
fire-places. This contrivance is very useful, especially for
closed fire-places that are without flues under the boilers,
as it occasions the flame to divide under the bottom of
the boiler, and to play over every part of it in a thin
sheet.

The reason why flues were not made under these
boilers was to render it possible to use occasionally

* For a particular account of these circular grates, see Vol. II, pp. 341 and
465, Plate I., Figs. 3 and 4. In Great Britain these grates may be made very
cheap of cast iron.

several boilers of different depths in the same fire-place; a convenience of no small importance in the kitchen of a private gentleman, who occasionally gives dinners to large companies.

It will be perceived that, in the fire-places of all the stewpans and saucepans, there are circular flues which oblige the flame to make one complete turn round the sides of the vessel, before it goes off into the horizontal canal; but I am far from being sure that the saving of fuel arising from this peculiar arrangement is sufficient to counterbalance the loss of that great convenience that results from being able to use indifferently stewpans and saucepans of different depths in the same stove, which cannot be obtained while these circular flues remain.

They will, indeed, be rendered unnecessary, provided that the flame be made to divide under the bottom of the vessel (which may be done by causing it to enter the horizontal canal by two opposite openings), and provided that this canal be furnished with a good damper, *which ought never to be omitted.* Although, to avoid the confusion that is apt to result from the delineation of a multitude of different objects in the same drawing, the dampers to the canals are all omitted in these plans, they must on no account be left out in practice, for they are of such importance that there is no possibility of managing fires properly without them; and as it is of very little importance whether they be placed near the fire or far from it, or what is their form, provided they be so constructed as to diminish at pleasure, and occasionally to close entirely the canal by which the smoke makes its escape, it is not necessary for me to give any particular directions how they are to be made; indeed, their construction is so very simple, and so generally known, that it would be quite superfluous for me to enlarge on that subject.

The dotted lines leading from the front of the brick-work to the fire-places show the position and dimensions of the ash-pits.

The whole length of the mass of brick-work from A to B is 11 feet, and its width from A to C is 7 feet 4 inches. The space it occupies on the ground may be conceived to consist of six equal squares of 44 inches each, placed in two rows of three squares each; these two rows being joined to each other by their sides, and forming together a parallelogram. And, in laying out the work when a kitchen is to be fitted up on the plan here described, it will always be best to begin by actually drawing these six squares on the floor of the kitchen. Nearly the whole of the middle square of the back row is occupied by the open chimney fire-place, and by its thick hollow walls; and the greater part of the middle square of the front row is left as a passage for the cook to come to the open chimney fire-place, or rather to the stoves that are situated near it.

PLATE IX.

Fig. 4. This figure, which represents a vertical section of the mass of brick-work through the centres of the fire-places of the four principal boilers, is chiefly designed to show the construction of those fire-places, and also that of the boilers. Sections of the circular grates on which the fires are made to burn under the boilers are here represented, and also sections of the ash-pits, and of the contractions of the fire-places immediately below the grates;*

* For an account of the utility of these contractions, see Vol. II, p. 343.

PLATE IX

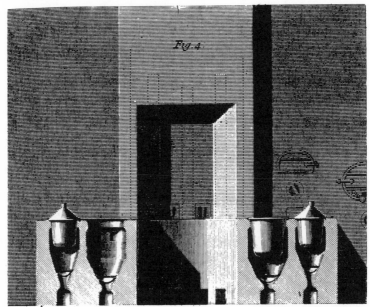

*Plans of Baron de Lerchenfelds Kitchen.*

and in one of the fire-places, which is shown without its
boiler, the openings of the branched canal by which
the smoke goes off horizontally towards the chimney
are also marked.

Fig. 5. This figure shows a bird's-eye view of the
upper surface of the brick-work, with all the boilers
and saucepans in their places, except one; three of the
principal boilers and one saucepan with their covers
on; and the rest of them without their covers. It
likewise represents a horizontal section of the open
chimney fire-place, 4 inches above the level of the top
of the mass of brick-work in which the boilers and
saucepans are set.

It is to be observed that all the boilers, stewpans, and
saucepans are fitted into circular rings of iron, which
are firmly fixed to the brick-work; and that they are
suspended in their fire-places by their circular rims.
All the stewpans and saucepans, that are not too large
to be lifted with their contents in and out of their fire-
places with the strength of one hand, have iron handles
attached to their circular rims; but the four principal
boilers, which are too large to be managed with one
hand, have each two rings fitted to their rims. These
handles and rings are so constructed that they do not
prevent the saucepans and boilers from fitting the
circular openings of their fire-places; neither do they
prevent their being fitted by their own circular covers.

It will, doubtless, be observed that the four principal
boilers shown in Fig. 4, belonging to the kitchen I am
now describing, differ but very little in form from the
boilers in common use, and consequently that they are
considerably deeper in proportion to their width than
they ought to be, in order that the heat generated in

the combustion of the fuel might act upon them to the greatest advantage; but it is to be remembered that to each of these fire-places there are other shallower boilers that are used occasionally, which do not appear in these plans. There is, however, one advantage attending deep boilers, to which it may in some cases be useful to pay attention; and that is, that they economize *space* in a kitchen. And when their fire-places are properly constructed, and, above all, when they are furnished with good registers and dampers, the additional quantity of fuel they will require will be too trifling to be considered. The walls of their fire-places will absorb more heat in the beginning; but who knows but that the greater part of this heat may not afterwards be emitted in rays, and at last find its way into the boiler? I could mention several facts that have lately fallen under my observation, which seem to render this supposition extremely probable. This, however, is not the proper place to give an account of them.

As I have said that no fire has yet been made in the open chimney fire-place of the kitchen I am describing, it may, perhaps, be asked how this kitchen is warmed in cold weather. To this I answer, that it has been found that the mass of brick-work is made sufficiently hot by the fires that are kept up in it when cooking is going on every day to keep the room comfortably warm in the coldest weather.

This answer will probably give rise to another question, which is, how we contrive to prevent the room from being much too warm in summer. By opening one of the windows a very little, and by opening at the same time the register of a wooden tube or steam-chimney, which, rising from the ceiling of the room, ends in the open air; and which is always opened to clear the room of vapour when it is found necessary, and especially when the victuals are taken out of the boilers, or when any other operation is going on that occasions the diffusion of a considerable quantity of steam. The oblong opening of this steam-chimney may be seen Plate VII., Fig. 1, in the ceiling, at the right-hand corner of the room.

Near this corner of the room may likewise be seen a front view of the hither end of one large roaster, and part of the front view of a smaller one situated by the side of it, both with their separate fire-place doors.

The fire-place door of the larger roaster, as also both its blowpipes, are represented as being open; but the ash-pit door of this roaster is hid by the mass of brick-work in which the boilers are set. A particular account of these roasters will be given hereafter.

The dimensions of the boilers in this kitchen are as follows : —

|  | Wide at the brim. Inches. | Deep. Inches. |
|---|---|---|
| One large boiler heated by smoke. . . . . . . . . . | 20 | 8 |
| Two large boilers . . . . . . . . . . . . . . | 16 | 16 |
| Two ditto, used occasionally in the fire-places of the two boilers last mentioned . . . . . . . . . . . . | 16 | 8 |
| Two smaller boilers . . . . . . . . . . . . . | 12 | 12 |
| Two ditto, fitted to the same fire-places . . . . . . . | 12 | 6 |

The diameters of the stewpans and saucepans are 12, 10, and 8 inches; and their depth is made equal to half their diameters.

The fuel burnt in this kitchen is wood; and the billets used are cut into lengths of about 6 inches.

Common bricks were used in the construction of the fire-places, but care was taken to lay them in mortar

PLATE X

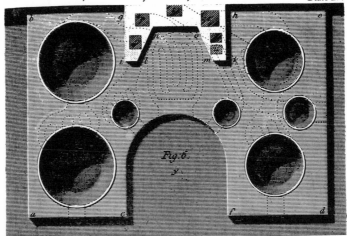

Kitchen of the Hospital of La Pieta at Verona.
Plate IV.

Fig. 6.

Kitchen of the House of Correction at Munich.

Fig. 7.

composed of clay and brickdust, without any sand, with only a very small proportion of lime.

In this kitchen, as also in that which I am now about to describe, the mass of brick-work in which the boilers are set projects into the room from the middle of one side of it.

### *Description of the Kitchen of the Hospital of La Pietà at Verona.*

#### PLATE X.

Fig. 6. This figure represents the ground plan of the mass of brick-work in which the boilers are fixed, and the canals by which the smoke is carried off from the fire-places into the chimney. The ground covered by this mass of brick-work, and by the area ($y$) between the boilers, may be conceived to be divided into six equal squares, of 43 inches, placed in two rows of three squares each. In the centres of four of these squares — namely, of those which are situated at the ends of the rows — are placed four large circular boilers. The middle square of the front row is chiefly occupied by the area which is left between the two front boilers; and one half of the middle square of the back row is occupied by an open chimney fire-place, in the thick walls of which no less than six vertical flues are concealed, which carry off the smoke from the boilers and stewpans into the chimney.

The smoke from the fire which heats the large boiler P (which boiler is $32\frac{1}{2}$ inches in diameter), on quitting its fire-place, goes off in four separate branches, which soon unite and form one canal, rises up under the middle of the bottom of the neighbouring large boiler Q, makes one complete turn under that boiler, and, passing from thence towards the centre of the mass of

brick-work, circulates in canals divided into several
branches under an iron plate that forms the bottom of
an oven, which is situated under the hearth of the open
chimney fire-place. From under the bottom of this oven
this smoke goes off obliquely, and, entering the bottom
of the vertical canal *p*, goes off into the chimney. The
principal use of this oven is to dry the wood that is used
as fuel in the kitchen. The large boiler Q, that is heated
by this smoke, is designed for warming water for the
use of the kitchen, and for various other purposes for
which hot water is occasionally used in the hospital.

The boiler P is principally used in preparing food for
the children in the hospital.

The smoke from the fire which heats the boiler R,
passing off in a canal which leads to the boiler S, there
separates, and passing round the sides of the boiler S,
and under a small part of its bottom, unites again, and
passes off into the chimney by the vertical canal *r*.
The heat in this smoke, though it is sufficient to *warm*
the water in the boiler S, is not sufficient to make it
boil. In order that the contents of this boiler may
occasionally be made boiling-hot, the boiler has a small
fire-place of its own, situated immediately under the
middle of its bottom; and when the water in the boiler
has been previously made warm by the smoke from the
boiler R, a very small fire made under it, in its own separate fire-place, will make it boil. The smoke from this fire-
place goes off by its own separate canal into the vertical
canal *s*, so that it does not interfere at all with the smoke
from the fire-place of the boiler R; and, in consequence
of this arrangement, the heating of the boiler S, by the
smoke from this neighbouring fire-place and by its own
fire, may be going on at the same time.

The smoke from the small boiler T, and from the stewpans U and W, goes off immediately by separate horizontal canals into their separate vertical canals (*t*, *u*, and *w*) that open into the chimney, at the height of about 15 inches above the mantel of the open chimney fire-place; and all the vertical canals, by which the smoke goes into the chimney, are furnished with dampers.

The side *b c* of the mass of brick-work is placed against the middle of one side of the kitchen, which is a large room; and the walls of the open chimney fire-place *g h i k* are carried up perpendicularly to the ceiling of the room. The hearth *l m n o* is on a level with the top of the brick-work in which the boilers are set.

As the principal boilers are deep, in order to provide sufficient room for them and a sufficient depth for their ash-pits, the foundation of the quadrangular mass of brick-work *a b c d* was raised 16 inches above the pavement of the kitchen; and on the three sides of the mass of brick-work *a b*, *a d*, and *d c*, which project into the room, there are two steps, 8 inches in height each, which extend the whole length of each of those sides; and for greater convenience in approaching the boilers the uppermost step is made 2 feet wide, and the area *y* is on a level with the top of this wide step. The ash-pit doors of the principal boilers are placed in the front of this step, and the bottoms of the passages or door-ways into their fire-places, by which the fuel is introduced, are situated just on a level with its upper surface.

The mass of brick-work in which the boilers are placed is 10 feet 9 inches long, and 8 feet 2 inches wide; and it is elevated to the height of about 3 feet 2 inches above the top of the upper broad step, by which it is surrounded on three sides, and on which it appears to stand.

## *Description of the Kitchen of the House of Correction at Munich.*

Plate X., Fig. 7, and Plate XI., Figs. 8 and 9, represent the plans and sections of this kitchen.

Fig. 7 represents the ground plan of the brick-work in which the boilers, etc., are set, or rather a horizontal section of the brick-work at the level of the fire-places, and of the canals for carrying off the smoke. In this kitchen the fires are not made on circular iron grates, as in that just described, but the fuel is burned on grates or bars composed of bricks set edgewise, as may be seen by the plans. (See *b*, *b*, *b*, etc., Fig. 7.)

The two principal boilers (*l*, *l*, Fig. 9) are quadrangular, each being 3 feet long, 2 feet wide, and 15 inches deep, furnished with wooden covers movable on hinges; and they are both heated by one fire. That which is situated in the front of the brick-work, and immediately over the fire, is used for making soup; while the other, which is placed very near it, and on the same level, is used for boiling meat, potatoes, greens, etc., in steam. A small quantity of water (about an inch in depth) being put into the second boiler, the smoke from the first, which passes in flues under the second, soon causes this water to boil, and fills the boiler with hot steam. The steam from the first boiler is also carried into the second by means of a tube about $\frac{3}{4}$ of an inch in diameter, furnished with a cock, which forms a communication between the two boilers just below the level of their brims. This tube of communication is not expressed in the plates.

The smoke, having quitted the second boiler, rises up obliquely to the level of the top of the mass of brick-work

PLATE XI

Front View of the Kitchen of the House of Correction at Munich

Fig.8.

Bird's-eye View.

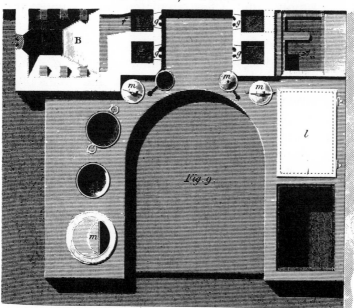

Fig.9.

in which the before-mentioned boilers are set, and then circulates under a quadrangular copper vessel (expressed by dotted lines at A, Fig. 8), 27 inches long, 19 inches wide, and 20 inches deep, destined for containing warm water for the use of the kitchen. As this vessel stands higher than the tops of the boilers, it is found to be very convenient for filling them with water; and, as this water is kept warm by the smoke, this arrangement produces a considerable economy of fuel as well as of time. The water is drawn off from this vessel for use by means of a brass cock, which is not expressed in the drawing; and it is supplied with water from a neighbouring reservoir, the entrance of the water being regulated by a regulating cock or valve, furnished with a swimming ball.

The smoke, after it has circulated in flues under this vessel, goes off into a vertical canal which conducts it into the chimney. This vertical canal, together with three others designed for a similar use (see *d, d, d, d*, Fig. 7, and Fig. 9), are situated in the thick walls of an open chimney fire-place (*n*, Fig. 8), the hearth of which is on a level with the top of the mass of brick-work in which the boilers are set. A horizontal section of these four vertical flues, taken at the height of 3 inches above the level of the hearth, and also a horizontal section of the brick-work of a roasting-machine (B, Figs. 8 and 9), situated on the left of this open chimney fire-place, are distinctly represented in the Fig. 9.

Under the hearth of the fire-place there is an open vault which serves as a magazine for fuel; and in the front wall of the fire-place, above the mantel, just under the ceiling of the room, there are two openings into the chimney, by which the steam that rises from the boilers escapes into the chimney, and goes off with the smoke.

The manner in which the flues are constructed under the different boilers, and the horizontal canal for carrying off the smoke from the round boilers into the chimney, are shown in the Fig. 7. The ash-pit doors to the two principal round boilers, which are expressed by dotted lines, are opposite to E and F, Fig. 7.

The ash-pit door belonging to the fire-place of the large quadrangular boilers is situated opposite to G, Fig. 7. The reason why these ash-pit doors were not placed immediately under their fire-place doors is because there was not room for them in that situation, owing to the pavement of the area between the boilers being raised one step higher than the floor of the kitchen, which was done for the convenience of the cook.

The openings for introducing the fuel into the fire-places are conical holes in square tiles, closed with earthen stoppers (see page 26). Though these tiles are not particularly distinguished in these plates, the stoppers which close their conical openings are shown. As these tiles are so worked into the mass of the brick-work as to make a part of it, and as they are plastered and white-washed in front, it is not easy to distinguish them from the bricks when the work is finished. Their joinings with the bricks in front could not therefore with propriety be marked in any of these plans.

Although the roaster belonging to the kitchen we are describing is not seen, yet the mass of brick-work in which it is fitted up appears on the left-hand side of the open chimney fire-place in Fig. 8; and a bird's-eye view of its fire-place, and of the projecting edges of the bricks on which it rests, is seen in the Fig. 9.

PLATE XII

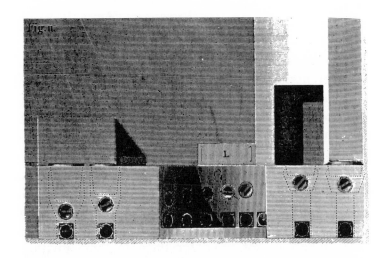

Plan & Elevation of the new Kitchen of the Military Hospital
at Munich.

## *Description of the new Kitchen in the Military Hospital at Munich.*

PLATE XII., FIGS. 10 AND 11, AND PLATE XIII., FIG. 12.

The mass of brick-work in which the boilers, the roaster, the stewpans, etc., are set, occupies one corner of the kitchen, extending 11½ feet on one side of the room, and 13 feet 7 inches on the other. The greatest width of the mass of brick-work (from A to B, or from C to D) is 50¾ inches, and its height from the floor 36 inches. The circular area (E, Figs. 9 and 10) in the angle of the mass of brick-work is 6 feet 8½ inches in diameter; and it is raised one easy step, or about 5 inches, above the level of the floor of the room. There is an open chimney fire-place of a peculiar form (F, Fig. 10) in the corner of this kitchen, the hearth of which is on a level with, or rather makes a part of the upper surface of, the mass of brick-work. The side-walls of this open chimney fire-place are hollow (see G and H, Fig. 10), and serve as canals for carrying off the smoke from the boilers into a chimney, which is situated quite in the corner of the room. These canals open into the chimney about 15 inches above the level of the mantel.

The smoke goes off from each fire-place by two separate and very narrow horizontal canals into larger common canals (see I and K, Fig. 9), which conducts it to the chimney; and the openings of these narrow canals are occasionally closed more or less by means of small pieces of brick or of earthen-ware, which serve instead of dampers, but which are not expressed in the plates. The fires all burn on flat grates, composed of bricks or thin tiles set edgewise. To save expense, the covers of the boilers and stewpans were all made of wood. The oblong quadrangular vessel (see L, Figs. 10 and 11), which is made of copper, and has a door above movable on hinges, is destined for containing warm water for the use of the kitchen, and is heated by the smoke from all the neighbouring closed fire-places.

The fire-place of the roaster is seen in Fig. 9 (M); a bird's-eye view of the top of the roaster appears in Fig. 10, and a vertical section of it and of its flues are faintly marked by dotted lines in Fig. 11.

The two large shallow stewpans (N, O, Fig. 10), vertical sections of which and of their fire-places are faintly marked by dotted lines in Fig. 11, are constructed of hammered iron, and are used principally for cooking steam dumplings (*dampf-nudels*), a kind of food in great repute in Bavaria.

When any thing is to be fried or broiled, a fire is made on the hearth of the open chimney fire-place. Under this hearth there is a small vault which serves for holding the wood that is wanted for fuel; but it would have been much better if that space had been occupied by two circular closed fire-places, so constructed as to be used occasionally for a frying-pan or a gridiron.

*Description of a detached Part of the Kitchen of the Military Academy at Munich.*

PLATE XIII.

Fig. 13. This figure is the ground plan of a mass of brick-work occupying a space about 6 feet 9 inches square, measured on the floor, in one corner of the

PLATE XIII

Plan of a Part of the Kitchen of the Military Academy.

Fig. 13.

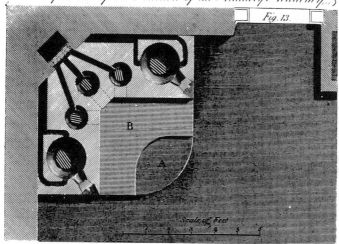

B

A

Scale of Feet

Fig. 12.

M

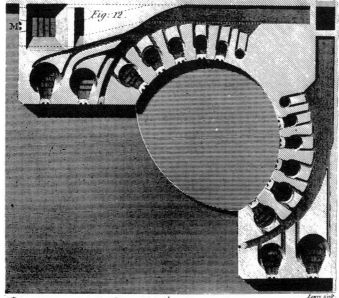

Lowry sculp.

Plan of the New Kitchen of the Military Hospital at Munich.

room, in which two of the principal boilers belonging to the kitchen, and three large stewpans, are fixed.

A and B are two steps, each 8 inches high; and the upper (flat) surface of the mass of brick-work, in which the boilers are set, and which is 45 inches wide, is just 30 inches above the level of the upper surface of the step B.

Neither the boilers nor stewpans are shown in this plan, but their circular fire-places are represented, as also their circular dishing iron grates, on which the fuel is burned, and the horizontal canals by which the smoke passes off into the chimney.

The smoke divides under each of the two principal boilers, and passes off in two canals situated on opposite sides of the fire-place; which canals, however, unite and form one single canal at a small distance from the boiler. In the fire-places of the stewpans the smoke does not divide in this manner; but the fire-place is so constructed that the flame makes one complete turn round the stewpan before it goes off into the horizontal canal leading to the chimney.

The opening by which the fuel is introduced into the fire-place of each of the two large boilers is closed by a conical stopper (constructed of fire-stone), represented in the figure, immediately under which stopper the (register) door of the ash-pit is situated.

The ash-pit of each of the fire-places of the stewpans is furnished with a register door. The passages into these ash-pits are expressed in the figure by dotted lines. The fuel (which is small pieces of wood about 5 inches in length) is introduced into the fire-place from above by removing the stewpan for a moment for that purpose.

The chimney C, by which the smoke goes off, is

situated in a corner of the room; and, when it is swept, the chimney-sweeper enters it by a door-way, which is situated in front, just above the level of the upper surface of the mass of brick-work, and which is closed by an iron door.

Each of the horizontal canals, by which the smoke is carried off from the fire-places of the two large boilers into the chimney, is furnished with a damper, which is faintly marked in the figure. Each of the horizontal canals, which carry off the smoke from the fire-places of the stewpans, is likewise furnished with a damper; but, to avoid confusion, they are not expressed in the engraving.

The bottoms of the ash-pit doors of the fire-places of the three stewpans are on a level with the upper surface of the step B; but the bottoms of the ash-pit doors of the fire-places of the two large boilers are on a level with the pavement of the kitchen.

The two large boilers (which are constructed of sheet copper, tinned) are 22 Rhinland inches in diameter above, $19\frac{1}{4}$ inches in diameter below, and 24 inches deep. They weigh each 62 lbs. avoirdupois, and contain 28 wine-gallons. The circular dishing-grates belonging to their fire-places are each 10 inches in diameter, measured externally; and the fire-place, properly so called, or the cavity in which the burning fuel is confined, is 10 inches in diameter below, 18 inches in diameter above, and $8\frac{1}{2}$ inches deep.

The largest stewpan is 12 inches in diameter, and 4 inches deep; and the two others are each 11 inches in diameter and 4 inches deep.

The fire-places belonging to the stewpans are cylindrical, 5 inches deep and 6 inches in diameter, and are furnished with circular dishing-grates.

Each of the large boilers is furnished with a circular wooden rim, 2 inches wide and 2 inches thick, which is accurately fitted to the brim of the boiler; and a circular wooden cover, consisting of three pieces of deal board attached to each other by two pairs of hinges, closes the boiler by being fitted accurately to the upper surface of its circular wooden rim.

One of the three pieces of board, which together form the flat circular cover of the boiler, is firmly fastened down to the wooden rim of the boiler, by means of two small hooks of iron; and from the middle of this part of the cover, so fastened down, a long tin tube, about $1\frac{1}{4}$ inches in diameter, rises up perpendicularly to the ceiling of the room, and carries off the steam from the boiler out of the kitchen.

As the cover of the boiler is composed of three flat pieces of board united by hinges, and as the cover, so formed, is merely laid down on the flat surface of the wooden rim which is connected with the brim of the boiler, it might very naturally be expected that some of the steam would be forced through between the joinings of the cover, or between the cover and the wooden rim; but this is what never happens. So far from it, steam seldom comes into the room even when the cover of the boiler is in part removed, by laying back the first division of it upon the second, — so strong is the draught of the steam-tube.

This phenomenon, which rather surprised me when I first observed it, was of considerable use to me; for it led me to discover the utility of dampers in the tubes or chimneys that are destined for carrying off the steam from boilers, and more especially from such boilers whose covers are not perfectly air-tight. If these steam-chimneys are of any considerable length, they cannot fail to occasion a strong draught through them, which will have a tendency to cause the cold air of the atmosphere to press in by every crevice between the brim of the boiler and its cover; which streams of cold air, being precipitated upon the surface of the boiling liquid, will be there warmed, and then passing off rapidly by the steam-chimney will occasion a very considerable loss of heat.

The rule for regulating the damper of the steam-chimney of a boiler, whose cover is not steam-tight, is this: close the damper just so much that closing it any more would cause some steam to be driven out between the joinings of the brim of the boiler and its cover. When this is done, it is evident that little or no cold air can enter the boiler by any small crevices in its cover that may remain open, consequently little or no heat will be carried off by the air of the atmosphere from the surface of the hot liquid.

I have been the more particular in explaining this matter, as I am persuaded that a great deal of heat is frequently lost in boiling and evaporating liquids, by causing or permitting the cold air of the atmosphere to come into contact with the surface of the hot liquid.

Some, I know, are of opinion that a stream of fresh air or a wind, which is made to pass over the surface of a liquid that is evaporated by boiling, tends rather to increase the evaporation than to diminish it; but it appears to me that there are strong reasons to conclude that this opinion is erroneous. A very simple experiment which I propose to make, and which others may perhaps be induced to make before I can find leisure to attend to it, will determine the fact.

The large boiler belonging to the fire-place, which is situated on the left hand in the mass of brick-work above described, is that which was used in the experiment mentioned on page 314 of Volume II.

It was once my intention to have published drawings and descriptions of every part and detail of the kitchen of the Military Academy at Munich, and also that of the House of Industry in that city. But as enough has already been said in this and in my sixth Essay to give clear and distinct ideas of the fundamental principles on which all the essential parts of the machinery in those kitchens were constructed; and as the peculiar arrangement of a kitchen must ever depend much on its size, and on the variety and kinds of food that are to be cooked in it, to avoid being tedious and tiresome to my readers, I have, after mature deliberation, concluded that it will be best to suppress these details.

Having now finished all the descriptions which I think it useful to publish of the various public and private kitchens that have been constructed under my direction in foreign countries, and having explained in the most ample manner in this Essay, and in my other writings on the management of fire, all the leading principles according to which, in my opinion, kitchens and fire-places of all kinds should be constructed, I shall in the next place proceed to show in what manner my plans may be so modified and accommodated to the opinions and practices in this country as to remove the objections that will probably be made to them, and facilitate their gradual introduction into general use.

I am well aware that it is by no means enough for those who propose improvements to the public to be in the right in regard to the intrinsic merit of their plans:

much must be done to prepare the way for, and to facilitate their introduction, or all their labours will be in vain.

———————

## CHAPTER III.

*Of the Alterations and Improvements that may be made in the Kitchen Fire-places now in common Use in Great Britain. — All Improvement in Kitchen Fire-places impossible, as long as they continue to be incumbered with Smoke-jacks. — They occasion an enormous Waste of Fuel. — Common Jacks, that go with a Weight, are much better. — Ovens and Boilers that are connected with a Kitchen Range should be detached from it, and heated each by its own separate Fire. — The closed Fire-places for iron Ovens and Roasters can hardly be made too small. — Of the various Means that may be used for improving the large open Fire-places of Kitchens. — Of the Cottage Fire-places now in common Use, and of the Means of improving them. — Of the very great Use that small Ovens constructed of thin sheet Iron would be to Cottagers. — Of the great Importance of improving the Implements and Utensils used by the Poor in cooking their Food. — No Improvement in their Method of preparing their Food possible without it. — Description of an Oven suitable for a poor Family, with an Estimate of the Cost of it. — Of Nests of three or four small Ovens heated by one Fire. — Of the Utility of these Nests of Ovens in the Kitchens of private Familïes. — They may be fitted up at a very small Expense. — Occa-*

THE kitchen fire-place of a family in easy circumstances in this country consists almost universally of a long grate, called a kitchen range, for burning coals, placed in a wide and deep open chimney, with a very high mantel. The front and bottom bars of the grate are commonly made of hammered iron, and the back of the grate (which usually slopes backwards) of a plate of cast iron; and sometimes there is a vertical plate of iron, movable by means of a rack in the cavity of the grate, by means of which plate the capacity, or rather the length of that part of the grate that is occupied by the burning fuel, may occasionally be diminished. At one end of the grate there is commonly an iron oven, which is heated by the fire in the grate; and sometimes there is a boiler situated in a similar manner at the other end of it. To complete the machinery (which in every part and detail of it seems to have been calculated for the express purpose of devouring fuel), a smoke-jack is placed in the chimney!

I shall begin my observations on the smoke-jack.

No human invention that ever came to my knowledge appears to me to be so absurd as this. A wind-mill is certainly a very useful contrivance, but were it proposed to turn a wind-mill by an artificial current of air, how ridiculous would the scheme appear! What an enormous force would necessarily be wasted in giving velocity to a stream of air sufficient to cause the mill to work with effect! A smoke-jack is, however, neither more nor less than a wind-mill, carried round by an

artificial current of air; and to this we may add that
the current of air which goes up a chimney, in consequence of the combustion of fuel in an open chimney
fire-place, is produced in the most expensive and dis-
advantageous manner that can well be imagined.    It
would not be difficult to prove that much less than *one
thousandth* part of the fuel that is necessary to be burned
in an open chimney fire-place, in order to cause a smoke-
jack to turn a loaded spit, would answer to make the
spit go round, were the force evolved in the combustion
of the fuel properly directed, — through the medium of
a steam-engine, for instance.

But it is not merely the waste of power or of mechanical force, that unavoidably attends the use of smoke-
jacks, that may be objected to them : they are very
inconvenient in many respects ; they frequently render
it necessary to make a great fire in the kitchen, when
otherwise a great fire would not be wanted ; they very
frequently cause chimneys to smoke, and always render
a stronger current of air up the chimney necessary than
would be so merely for the combustion of the fuel wanted
for the purposes of cooking; consequently they increase
the currents of cold air from the doors and windows to
the fire-place ; and, lastly, they are troublesome, noisy,
expensive, frequently out of order, and never do the
work they are meant to perform with half so much
certainty and precision as it would be done by a common jack, moved by a weight or a spring.

There is, I know, an objection to common jacks that
is well founded, which is, that they require frequent
winding up; but for this there is an easy remedy.    A
jack may without any difficulty (merely by using a
greater weight and a greater combination of pulleys)

be made to run almost any length of time: a whole day for instance, or even longer; and, if it should be necessary, the weight may be at a considerable distance from the kitchen. It may indifferently be raised up into the air, descend into a well, or may be made to descend along an inclined plane; and but little ingenuity will be required to contrive and dispose of the machinery in such a manner as to keep it out of the way, and, if it should be required, completely out of sight; and, with regard to the winding up of such a jack as I here recommend (that is, to go a whole day), it may easily be done by any servant of the house in less than five minutes.

Incomparably less labour will be required to wind up the weight of a common jack than to bring coals to feed the fire that is requisite to make a smoke-jack go.

I know that it is said in favour of smoke-jacks, that all the fire that is required to make them perform would be necessary in the kitchen for other purposes, and consequently that they occasion no additional expense of fuel; but that this statement is very far indeed from being accurate will be evident to any person who will take the trouble to examine the matter with care. That the sails of a smoke-jack will turn round with the application of a very small force, when the pivots on which its axle-tree rests are well constructed, and when its motion is not impeded by any load, is very true; but it requires a very different degree of force to move it when it is obliged to carry round one, or perhaps two or three, loaded spits. Even the heat given off to the air by the kitchen range in cooking, after the fire is gone out, will sometimes keep up the motion of the sails of the smoke-jack for many hours. But what a striking proof is this of the enormous waste of fuel in kitchens in this country!

Would to God that I could contrive to fix the public attention on this subject.

Nothing surely is so disgraceful to society and to individuals as unmeaning wastefulness.

But to return to the attack of my smoke-jack; which (although it be a *wind-mill*) is certainly not a *giant*, and cannot be personally formidable, however it may expose me to another species of danger.

There is one objection to smoke-jacks that must be quite conclusive wherever the improvements I have recommended, and shall recommend, in kitchen fire-places, are to be introduced. Where smoke-jacks exist, these improvements cannot be introduced, it being quite impracticable to unite them.

On a supposition that I have gained my point, and that the smoke-jack is to be removed, I shall now proceed to propose several alterations and improvements that may be made in the kitchen range.

And, first, all ovens, boilers, steam-boilers, etc., which are connected with the back and ends of the range, and heated by the fire made in the grate, should be detached from it; and for each of the ovens, boilers, etc., a small, separate, closed fire-place must be constructed, situated directly under the oven or boiler, and furnished with a separate canal for carrying its smoke into the kitchen chimney, which separate canal may open into the chimney about a foot above the level of the mantel.

There is nothing so wasteful as the attempt to heat ovens and boilers by heat drawn off laterally from a fire in an open grate. The consumption of fuel is enormous, to say nothing of the expense of the machinery, and the inconvenience that must frequently arise from the heat being forcibly drawn away sidewise under an oven or boiler, when it is wanted elsewhere.

The separate closed fire-place under iron ovens and roasters must be made *very small*, otherwise the cook or his assistants will sometimes, in the hurry of business, make too large a fire; the consequences of which will be the spoiling of the food, and the burning and destroying of the oven or roaster.

Almost all the roasters that have been put up in England have been spoiled in consequence of their fire-places being made too large; and not one has ever received the slightest accident or injury, or failed to perform to entire satisfaction, that has been heated by a very small fire, and never overheated.

The fire-place for an oven or roaster of sheet iron, from 18 to 20 inches wide, and from 24 to 30 inches long, should never be more than 6 inches wide, 6 inches deep, and about 9, or at most 10, inches long; and this fire-place should seldom be half filled with coals. If the oven or roaster be set in such a manner that the flame or smoke from the fire must necessarily spread round it and embrace it on every side, there will be no want of heat for any of the common purposes of cookery, and its intensity may at all times be regulated by means of the damper in the chimney and the register in the ash-pit door.

It is not easy to imagine how much the business of cooking is facilitated by making the machinery so perfect that the quantity of heat may at any time be regulated with certainty merely by registers and dampers, and without adding to or diminishing the quantity of fuel in the fire-place. It is on these advantages, and the numerous other conveniences that will result from them, that my hopes are principally founded of gaining over the cooks, and engaging their cordial assistance in bringing forward into general use the improvements I recommend. I am well aware of their influence, and of the importance of their co-operation.

When all the ovens and fixed boilers are detached from the kitchen range, then, and not before, measures may be taken with some prospect of success for improving the kitchen fire-place, so as to economize fuel, and prevent the kitchen chimney from smoking, if it has that fault; and the measures proper to be adopted for obtaining those ends must depend principally on the size, or rather on the width, of the open fire that will be wanted in the kitchen. Where the family is small, and where great dinners are seldom or never given, and especially where closed roasters are introduced, a small fire-place, and consequently a narrow grate, will answer every purpose that can be wanted; and the fire-place of the kitchen may be fitted up nearly upon the principles laid down in my fourth Essay,[6] on the construction of open chimney fire-places.

The kitchen of Mr. Summers, ironmonger, of New Bond Street (No. 98), has been fitted up in this manner, and has been found to answer perfectly well.

But if it be necessary to leave the grate of the kitchen range with its width undiminished, in order that a wide fire may occasionally be lighted in it, this can best be done in the manner that was lately adopted in altering and fitting up the kitchen in the house of the Countess of Morton in Park Street. The range being suffered to remain (or rather the front and bottom bars of the grate only, for the iron plate that formed the back of the range was taken away), the range, which is about 5 feet long, was divided into three unequal parts, which parts were built up with hard fire-bricks in such a manner as to form three distinct fire-places, the one contiguous to the other, and separated from each other by divisions so thin in front that when fires are burning in them all it appears like one fire, and has all the effect of one fire in roasting meat that is put before it. Each fire-place is, however, perfectly distinct from the others, and has its own distinct coverings (which are oblique), — back, throat, etc., — though the same front bars, which are of hammered iron, and made very strong, run through them all.

When a very small fire is wanted (merely for boiling a tea-kettle, for instance), it is kindled in the *first* or smallest fire-place; when a little larger fire is necessary, it is made in the *second* fire-place, which is at the opposite end of the range; when a still larger fire is required, it is made in the *third* fire-place, which occupies the middle of the range. If a large fire in the fourth degree is wanted, two neighbouring fires are kindled in the *first* and *third* fire-places; if in the fifth degree, the two contiguous fires are lighted in the second and third fire-places; and when the greatest fire that can be made is wanted, all the three fire-places are at the same time filled with burning fuel.

In cases where a single open chimney fire-place of a moderate size, that is to say, from 18 to 20 inches in width, might sometimes be too small, and a very wide fire, like that just described, would never be wanted, I would advise the construction of two separate but adjoining fire-places, the one about 12 inches, and the other about 18 or 20 inches in width. These would, I imagine, answer every purpose for which an open fire in the kitchen could be wanted by a large family, even though they should (contrary to all my recommendations) continue to roast their meat upon a spit.

That I am not unreasonable enough to expect that all my recommendations will immediately be attended to, is evident from the pains I take to improve machinery now in use, of which I do not approve, and which is perfectly different from that I am desirous to see introduced.

When my roasters shall become more generally known, and the management of them better understood, I have no doubt but that open chimney fire-places, and open fires of all descriptions, will be found to be much less necessary in kitchens than they now are.

I am even sanguine enough to expect that the time will come when open fires will disappear, even in our dwelling-rooms and most elegant apartments. Genial warmth can certainly be kept up, and perfect ventilation effected much better without them than with them; and though I am myself still child enough to be pleased with the brilliant appearance of burning fuel, yet I cannot help thinking that something else might be invented equally attractive to draw my attention and amuse my sight, that would be less injurious to my eyes, less expensive, and less connected with dirt, ashes, and other unwholesome and disagreeable objects.

It is very natural to suppose that those nations who inhabit countries where the winter is most severe must have made the greatest progress in contriving means for making their dwellings warm and comfortable in cold weather; and when, in milder climates, the growing scarcity of fuel has rendered the saving of that article an object of rational economy, it appears to me to be wise to search *there* for the means of doing it, where necessity has long since rendered the use and highest possible improvement of those means indispensable. And the truly liberal — that is to say, the enlightened, just, and generous — feel no difficulty in acknowledging the ingenuity and industry of their neighbours, and no humiliation in adopting their useful inventions and improvements.

Before I finish this publication I must say a few words on the construction of *cottage fire-places*. It is, I am sensible, a long time since I promised to publish an Essay on that subject, and still mean to do so; but a variety of weighty considerations has engaged me to postpone the putting of that Essay out of my hands. I conceived the subject to be of very great importance, and wished to have time to make myself fully acquainted with the present state of cottages, and of the different kinds of fuel used in them in different parts of these kingdoms. I had with pain observed the numerous mistakes that have been made in altering chimney fire-places on the principles recommended in my fourth Essay, and on that account I was very desirous of deferring the publication of my directions for constructing cottage fire-places, till I could inform the public where cottage fire-places, constructed on the principles recommended, might be seen.

I hope and trust that in the arrangement of the repository of the Royal Institution, now fitting up in this metropolis, an opportunity will be found for exhibiting cottage fire-places on the most perfect plans, as also of showing many other mechanical contrivances that may be of general utility.

Cottage chimneys, as they are now commonly constructed in most parts of Great Britain, have a very
wide open fire-place, with a high mantel, and large
chimney-corners, in which the children frequently sit
on little stools, when in cold weather they hover round
the fire.    These chimney-corners are very comfortable;
and, except the whole room could be made equally so,
it would certainly be a pity to destroy them.    But this,
I am persuaded, may easily be done: in the mean
time, much may be done to make cottages warm and
comfortable, merely by a few simple alterations in their
present fire-places.

As the principal fault of these fire-places is the
enormous width of the throats of their chimneys, which
frequently occasions their smoking, and always gives
too free a passage for the warm air of the room to
escape up the chimney, a smaller fire-place may be
constructed in the midst of the larger one; and the
little chimney of this small fire-place being carried up
perpendicularly in the middle of the large fire-place,
the large chimney-corners, without being destroyed,
may be arched over and closed in above, so as to leave
no passage in those parts for the escape of the warm
air of the room into the chimney, and from thence into
the atmosphere.

The back of the old chimney may serve for a back to
the new fire-place, and the jambs of the new chimney
need not project forward beyond the back more than
12 or 15 inches; so that the new chimney, and every
part of it, may be completely included within the
opening of the old fire-place.    This is to be done in
order to preserve the old chimney-corners; but in cases
where the opening of the old fire-place is not sufficiently

wide, high, and deep to permit of the leaving of chimney-corners sufficiently spacious to be useful, it will be best to sacrifice these corners, and to proceed in a different manner in constructing the new fire-place.

In this last case the back of the new fire-place should be brought forward, and the new work should be executed agreeably to the directions contained in my fourth Essay for the construction of open chimney fire-places. If void spaces should remain on the right and left of the new jambs, they will be found useful for various purposes.

It is of so much importance to facilitate the means of cooking to the poor, and to enable them to prepare food in different ways, that I think it extremely desirable that each cottager should have an iron pot or digester, so contrived as to be used occasionally over his open fire, or, what will be much more economical, in a small closed fire-place, which may be made with a few bricks on one side of his open fire-place.

But what would be of more use, if possible, to a poor family, even than a good boiler, would be a small oven of sheet iron, well put up in brick-work. Such an oven would not cost more than a few shillings, and if properly set would last for many years without needing any repairs. It would answer not only for baking household bread and cakes, but might likewise be used with great advantage in cooking rice puddings, potato pies, and many other kinds of nourishing food of the most exquisite taste, that might be prepared at a very trifling expense.

It is in vain to expect that the poor should adopt better methods of choosing and preparing their food, till they are furnished with better implements and utensils for cooking.

I put up an oven like that I now recommend last winter in my lodgings at Brompton, and have made a great number of experiments with it, from the results of which I am fully persuaded of its utility. I pulled it down on removing into the house I now occupy, but mean to put it up again as soon as my kitchen shall be ready to receive it. As I put up this oven merely as an experiment, in order to ascertain by actual trials how far it might be useful to poor families, the oven was made small, and it was set in the cheapest manner, merely with common bricks and mortar, without any iron or other costly material. The grate of the closed fire-place (which was 5 inches wide and about 8 inches long) was constructed of three common bricks placed edgewise, and a sliding brick was used for closing the door of the fire-place, and another for a register to the ash-pit door-way. The oven, which is of thin sheet iron, is $18\frac{1}{2}$ inches long, 12 inches wide, and 12 inches high, and it weighs just $10\frac{1}{2}$ lbs. exclusive of its front frame and front door, which together weigh $6\frac{1}{4}$ lbs.

For a small family the oven might be made of a smaller size, — 11 inches wide, for instance, 10 inches high, and 15 inches long; and it is not indispensably necessary that it should have either a front frame or a front door of iron. It might be set in the brick-work without a frame perfectly well; and a flat twelve-inch tile, or a flat piece of stone, or even a piece of wood, placed against its mouth, might be made to answer instead of an iron door.

The only danger of injury to these ovens from accident to which they are liable is that arising from carelessness in making too large a fire under them. They require but a very small fire indeed, and a large one is not only quite unnecessary, but detrimental on several accounts.

For greater security against accidents from too strong fires, I would advise the fire-place to be made extremely — I had almost said ridiculously — small, not more than from 4 to 5 inches wide, from 6 to 8 inches long, and about 5 inches deep; and I would place the bottom or grating of the fire-place 11 or 12 inches below the bottom of the oven. For still greater security, the bottom of the oven, immediately over the fire, might, if it should be found necessary, be defended by a thin plate of cast, hammered, or sheet iron, full of small holes (as large as peas), placed about half an inch from the bottom of the oven, and directly below it; but, if any common degree of attention be used in the management of the fire, this precaution will not, I am persuaded, be necessary.

In setting these ovens, care must be taken that room be left for the flame and smoke to come into contact with the oven, and surround it on every side; and it can hardly be necessary to add that a canal must be made by which the smoke can afterwards pass off into the chimney.

I once imagined that small ovens for poor cottagers might be made very cheap indeed, by making only the bottom of the oven of iron, and building up the rest with bricks; but, on making the experiment, it was not found to answer. I caused several ovens on this principle to be constructed in my kitchen, and made many attempts to correct their faults; but I found it impossible to heat them equally and sufficiently. I then altered my plan, by making both the bottom and the top of sheet iron. But this even did not answer. It might answer for a perpetual oven, like that which I caused to be made in the House of Industry at Dublin; but, if an oven of this kind is ever suffered to become cold, it will require a long time to heat it again, which is a circumstance that renders it very unfit for the use of a poor family. The ovens I have recommended, constructed entirely of thin sheet iron, have the advantage of being heated almost in an instant; and the heat which penetrates the walls of their closed fire-places, being gradually given off after all the fuel is burned out, keeps them hot for a long time. Care should, however, always be taken to keep these ovens well closed when they are used, and to leave only a very small hole, when necessary, for the escape of the generated steam or vapour.

For larger families the oven may be made larger in proportion; or, what will be still more convenient, a nest of two, three, or four small ovens, placed near to each other, may be so set in brick-work as to be heated by one and the same fire.

A nest of four small ovens, set in this manner, was fitted up in the kitchen of the Military Academy at Munich, and found very useful: they were rectangular, each being 10 inches wide, 10 inches high, and 16 inches long; and they were placed two abreast in two rows, one immediately above the other, the sides and bottoms of neighbouring ovens being at the distance of about $1\frac{1}{2}$ inch, that the flame and smoke which surrounded them on every side might have room to pass between them. The fire-place was situated immediately below the interval that separated the two lowermost ovens, at the distance of about 10 inches below the level of their bottoms; and by means of dampers the flame could be so turned and directed as to increase or diminish the heat in any one or more of the ovens at pleasure.

These four ovens were furnished with iron doors, movable on hinges, which, in order that they might not be in the way of each other, opened two to the right, and two to the left.

In a large kitchen, where a variety of different kinds of food is baked at the same time or on the same day, it is easy to perceive that a nest of small ovens must be very useful, much more so than one large oven equal in capacity to them all; for, besides the inconvenience in cooking a variety of different things in the same oven that arises from the promiscuous mixture of various exhalations and smells, the process going on in one dish must often be disturbed by opening the oven to put in or take out another, and the heat can never be so regulated as to suit them all.

But the cook of the Military Academy at Munich finds the nest of ovens useful not merely for baking: he uses them also for stewing and for boiling, with great success. A large quantity of cold liquid cannot, it is true, be heated and made to boil in a very short time in one of these ovens; but a saucepan or boiler, whose contents are already boiling-hot, being placed in one of them, a gentle boiling may be kept up for a great length of time, with the consumption of an exceedingly small quantity of fuel.

With regard to the expense or cost of such a nest of ovens, it could not, or at least ought not to, be considerable. If they were each 12 inches wide, 12 inches high, and 16 inches long, they would not weigh more than 15 lbs. each, their doors included; and this would make but 60 lbs. for the weight of the whole nest, supposing it to consist of four ovens. I do not know what price might be demanded by the artificers in this country, or by the trade, for work of this kind, but I should think they might well afford to sell these ovens,

properly made and ready for setting, at less than 6*d.*
the pound, avoirdupois weight. The sheet iron would
cost them in the market, at the first hand, not more than
about 3½*d.* per pound. The expense of setting the ovens
would not be considerable, especially as only one small
fire-place would be necessary.

In some future publication, or in a subsequent part
of this Essay, I shall give a design of one of these
nests of ovens, with an exact estimate of the expense of
it: in the mean time I will endeavour to get one of
them put up for the public inspection at the Royal
Institution.

I cannot close this chapter without once more calling
the attention of my reader to the necessity of furnishing the canal that carries away the smoke into the
chimney with a damper. If this is not done in setting
the ovens I have just been describing, it will be quite
impossible to manage the heat properly. For the fire-
place of a small oven for the family of a cottager, a
common brick may be made to answer very well as a
damper; and, indeed, a very good damper for any small
fire-place may be made with a brick or a tile or a
piece of stone.

If, in addition to the introduction of a good damper,
care be taken to cause the smoke to descend about
12 or 15 inches just after it has quitted the oven (or the
boiler), and before it is permitted to rise up and go off
into the chimney, this will greatly contribute to the
economy of fuel.

It is surely not necessary that I should again observe
how very essential it is in altering open chimney fire-
places — whether they belong to kitchens, to the dwelling-rooms of the opulent, or to cottages — to build up

their backs and sides, in that part especially which contains and is occupied by the burning fuel, with fire-bricks or with stone; and never in any case to kindle a fire against a plate of iron.

If all the metal in a register stove, except the front, and the front and bottom bars, were removed, and the back and sides built up properly with fire-bricks, or partly with fire-bricks and partly with fire-stone, it would make a most excellent fire-place.

This last observation is, I acknowledge, in some degree foreign to my present subject; but, as it is well meant, I hope it will be well received.

In a supplementary Essay now preparing for the press, in which will be published such additional remarks and observations to all my former Essays as may be necessary to their complete explanation and elucidation, I shall take occasion to enter fully into the subject of chimney fire-places, and shall endeavour to show, at some length, why it is improper and ill-judged to construct the sides and backs of their grates of iron, or of any other metallic substance.

In a second part which will be added to this (tenth) Essay, particular directions will be given for constructing boilers, steam dishes, ovens, roasters, and various other implements and utensils used in cookery; and a detailed plan will be laid before the public for improving the kitchen utensils of cottagers and other poor families.

I have been induced to reserve these various matters for a separate publication, in order to accommodate my writings as much as is possible to the convenience of the various classes of readers into whose hands they are likely to come. The plates, which were indispensably necessary to elucidate the descriptions contained in the preceding chapters (which have been admirably executed by that excellent artist Lowry), could not fail to enhance very considerably the price of this publication, and on that account I was desirous to detach and publish separately all such popular parts of the subjects I have undertaken to treat in this Essay as appeared to me to bid fair to be most read, and to be of most general utility.

Whether the reader agrees with me or not in respect to the validity of the reasons which have determined my judgment on this occasion, I hope and trust that he will do me the justice to believe that I have no wish so much at my heart as to render my labours of some real and lasting utility to mankind. How happy shall I be when I come to die, if I can *then* think that I have lived to some useful purpose!

# APPENDIX TO PART I.

---

*An Account of the Expense of fitting up a small Oven.*

SINCE the foregoing sheets were printed off, I have caused a small oven of sheet iron to be made and set in brick-work, for the express purpose of ascertaining the cost of it. This oven, which is such as would be proper for the use of a small poor family, is 11 inches wide, 11 inches high, and 15¾ inches long; and it weighs 6 lbs. 2 oz. At its mouth or opening, the sheet iron is turned back in such a manner as to form a rim, half an inch wide, projecting outwards; which rim serves to strengthen the oven, and is likewise useful in fixing it in the brick-work.

The whole oven is constructed of two pieces of sheet iron, of unequal dimensions, the largest piece (which is about 16½ inches wide by 45 inches long) forming the top, bottom, and two sides; and the smallest (which is about 12 inches square) forming the end. These sheets of iron are united by seams without rivets. One seam only runs through the oven in the direction of its length, and that is situated in the middle of the upper part of it.

A good workman was employed just two hours in making this oven; but there is no doubt but the work might be done in a shorter time by a man accustomed to that kind of manufacture, especially if the proper

141

means were used for facilitating and expediting the labour.

The sheet iron used in the construction of this oven, which was of the very best quality, cost 34*s.* per gross hundred of 112 lbs., which is at the rate of $3\frac{1}{2}d.$ and $\frac{3}{14}$ of a farthing per lb. The quantity used, 6 lbs. 2 oz., must therefore have cost 1*s.* $10\frac{1}{2}d.$ and $\frac{1}{112}$ part of a farthing.

If now we allow two ounces for wastage, this will bring the quantity necessary for constructing one of these ovens to $6\frac{1}{4}$ lbs., which quantity, at the rate above mentioned, would cost something less than 1*s.* 11*d.*; and if to this sum we add 1*s.* for the making, this will bring the prime cost of the oven to 2*s.* 11*d.*

Let us allow 20 per cent for the profit of the manufacturer, and still the price of the oven to buyers will be only 3*s.* 6*d.**

In order to ascertain the expense of setting one of these ovens in brick-work, I caused that above described to be put up in the middle of a wide chimney fire-place in my house in Brompton Row; and the work was executed with as much care and attention as was necessary, in order to render it strong and durable. In doing this 114 bricks were used, and something less than 3 hods of mortar; and the bricklayer performed the job in 3 hours and 10 minutes.

Three bricks set edgewise formed the grate or bottom of the fire-place; the middle brick being placed vertically, and those on each side of it inclining a little

---

* The oven I have here described was made by Mr. Summers, ironmonger, of New Bond Street, who, before I acquainted him with the above computations, offered to furnish these ovens in any quantities at 4*s.* a piece. This, for the offer of a manufacturer, I thought not unreasonable.

inwards above, to give a more free passage to the falling ashes.

The entrance into the fire-place was closed with a sliding brick, and another brick served as a register to the ash-pit door-way; a third served as a damper to the canal that carried off the smoke into the chimney ; and the oven itself was closed with a twelve-inch tile.

The expense of setting this oven was estimated as follows : —

|  | s. | d. |
|---|---|---|
| 114 bricks, at 3s. per hundred . . . . . . . . | 3 | 4 |
| 3 hods of mortar, at 4d. . . . . . . . . . . | 1 | 0 |
| 1 twelve-inch tile, at 4d. . . . . . . . . . . . | 0 | 4 |
| Bricklayer's labour . . . . . . . . . . . | 1 | 6 |
| Total . . . . . . . . . . . . . | 6 | 2 |
| If to this sum we add the amount of the ironmonger's bill for the oven . . . . . . . . . . . . . . | 3 | 6 |
| The whole expense will turn out . . . . . . . . | 9 | 8 |

The mass of brick-work in which this oven is set is just 2 feet wide, 19½ inches deep, measured from front to back, and 3 feet 3½ inches high. The chimney fire-place in which it is placed is 3 feet wide, 3 feet 3½ inches high, and 20 inches deep.

If the oven had been set in one corner of this fire-place, instead of occupying the middle of it, near one-quarter of the bricks that were used might have been saved; but if in building a new chimney a convenient place were chosen and prepared for it, an oven of this kind might be put up at a very small expense indeed, perhaps for 3s. or 3s. 6d., which would reduce the cost of the oven when set to about 7s. or 7s. 6d.

Though the bricklayer was above 3 hours putting up this oven, yet, as it was the first he ever set, there is no doubt but that he was considerably longer in doing the work on that account.   He thinks he could put up another in two hours, and I am of the same opinion.

I think it would be advisable, in order to facilitate stowage and carriage of these small ovens, always to manufacture them in nests of four, one within the other, even when they are designed to be sold, and to be put up singly; for it can be of no great importance whether they be a quarter of an inch or half an inch wider or narrower; and it will often be a great convenience to be able to pack them one within the other, especially when they are to be sent to any considerable distance.

If care be taken in making them to preserve their forms and dimensions, and if the seams of the metal be properly beaten down, the difference in the sizes of two ovens that will fit one within the other need not be very considerable.   But I forget that I am writing for the cleverest and most experienced workmen upon the face of the earth, to whom the utility of these contrivances is perfectly familiar, and who, without waiting for my suggestions, will not fail to put them all in practice.

Though there is nothing I am more anxious to avoid than tiring my reader with useless repetitions, yet I cannot help mentioning once more the great importance of causing the smoke that heats one of the ovens I have been describing to descend at least as low as the level of the bottom of the oven, after it has passed round and over it, before it is permitted to rise up freely and escape by the chimney into the atmosphere.   In setting the oven, and forming the canal for carrying off the smoke from the oven into the chimney, this may easily be effected: and, if it be done, the oven will retain its heat for a great length of time even after the fire is gone out; but, if it be not done, the fire must constantly be kept up, or the oven will soon be cooled by the cold air that will not fail to force its way through the fire-place and up the chimney.

From the result of this experiment it appears that an oven of the kind recommended is very far from being an expensive article; and there is no doubt but that, with a little care in the management of the fire, an oven of this sort would last many years without wanting any repairs. It is hardly necessary for me to add that a nest of these small ovens, consisting of three or four, put up together, and heated by a single fire, would be very useful in the kitchen of a private gentleman, and indeed of every large family.

If nests of small ovens should come into use (which I cannot help thinking will be the case), it would be best, as well for convenience in carriage as for other reasons, to make those which belong to the same nest not precisely of the same dimensions, but varying in size just so much as shall be necessary in order that they may be packed one within the other.

# PART II.

## PREFACE.

I TOO often find myself in situations in which I feel it to be necessary to make apologies for delays and irregularities in the publication of my writings. This second part of my tenth Essay was announced in the beginning of the year 1800; and it ought certainly to have made its appearance long ago, but a variety of circumstances has conspired to retard its publication.

During several months, almost the whole of my time was taken up with the business of the Royal Institution; and those who are acquainted with the nature and objects of that noble establishment will, no doubt, think that I judged wisely in preferring its interests to every other concern. For my own part, I certainly consider it as being by far the most useful, and consequently the most important, undertaking in which I was ever engaged, and of course I feel deeply interested in its success. The distinguished patronage and liberal support it has already received afford good ground to hope that it will continue to prosper, and be a lasting monument of the liberality and enterprising spirit of an enlightened nation.

It is certainly a proud circumstance for this country that in times like the present, and under the accumulated pressure of a long and expensive war, individuals generously came forward and subscribed in a very short time no less a sum than *thirty thousand pounds sterling,* for the noble purpose of " diffusing the knowledge and facilitating the general introduction of new and useful inventions and improvements."

In the *repository* of this new establishment will be found specimens of all the mechanical improvements which I have ventured to recommend to the public in my Essays.

# CHAPTER IV.

*An Account of a new Contrivance for roasting Meat.
— Circumstance which gave rise to this Invention.—
Means used for introducing it into common Use.
— List of Tradesmen who manufacture Roasters.—
Number of them that have already been sold. — De-
scription of the Roaster. — Explanation of its Action.
—Reasons why Meat roasted in this Machine is better
tasted and more wholesome than when roasted on a
Spit. — It is not only better tasted, but also more in
Quantity when cooked. — Directions for setting Roast-
ers in Brick-work. — Directions for the Management
of a Roaster. — Miscellaneous Observations respecting
Roasters and Ovens.*

THERE is no process of cookery more troublesome
to the cook, or attended with a greater waste of
fuel, than roasting meat before an open fire.

Having had occasion, several years ago, to fit up a
large kitchen (that belonging to the Military Academy
at Munich) in which it was necessary to make arrangements for roasting meat every day for near 200 persons,
I was led to consider this subject with some attention;
and I availed myself of the opportunity which then
offered to make a number of interesting experiments,
from the results of which I was enabled to construct
a machine for roasting, which upon trial was found
to answer so well that I thought it deserving of being
made known to the public. Accordingly, during the

148

visit I made to this country in the years 1795 and 1796, I caused two of these roasters to be constructed in London,—one at the house then occupied by the Board of Agriculture, and the other at the Foundling Hospital; and a third was put up, under my direction, in Dublin, at the house of the Dublin Society.

All these were found to answer very well, and they were often imitated; but I had the mortification to find, on my return to England in the year 1798, that some mistakes had been made in the construction, and many in the management of them. Their fire-places had almost universally been made three or four times as large as they ought to have been, as neither the cooks, nor the bricklayers who were employed in setting them, could be persuaded that it was possible that any thing could be sufficiently roasted with a fire which to them appeared to be *ridiculously small;* and the large quantities of fuel which were introduced into these capacious fire-places not only destroyed the machinery very soon, but, what was still more fatal to the reputation of the contrivance, rendered it impossible for the meat to be well roasted.

When meat, surrounded by air, is exposed to the action of very intense heat, its surface is soon scorched and dried; which preventing the heat from penetrating freely to the centre of the piece, the meat cannot possibly be equally roasted throughout.

These mistakes could not fail to discredit the invention, and retard its introduction into general use; but, being convinced by long experience of the utility of the contrivance, as well as by the unanimous opinion in its favour of all those who had given it a fair trial, I was resolved to persist in my endeavours to make it

known, and, if possible, to bring it into use in this
country. The roaster in the kitchen of the Military
Academy at Munich had been in daily use more than
eight years; and many others in imitation of it, which
had been put up in private families in Bavaria and
other parts of Germany, and in Switzerland, had been
found to answer perfectly well; and as that in the
kitchen of the Foundling Hospital in London had
likewise, during the experience of two years, been found
to perform to the entire satisfaction of those who have
the direction of that noble institution, I was justified in
concluding that, wherever the experiment had failed, it
must have been owing to mismanagement. And I was
the more anxious to get this contrivance brought into
general use, as I was perfectly convinced that meat
roasted by this new process is not merely as good, but
decidedly better; that is to say, more delicate, more
juicy, more savoury, and higher flavoured, than when
roasted in the common way, — on a spit, before an
open fire.

A real improvement in the art of cookery, which
unites the advantage of economy with wholesomeness,
and an increase of enjoyment in eating, appeared to me
to be very interesting; and I attended to the subject
with all that zeal and perseverance which a conviction
of its importance naturally inspired.

On my return to this country, in the autumn of the
year 1798, one of the first things I undertook in the
prosecution of my favourite pursuit was to engage an
ingenious tradesman, who lives in a part of the town
which is much frequented (Mr. Summers, ironmonger,
of New Bond Street), to put up a roaster in his own
kitchen; to instruct his cook in the management of it;

to make daily use of it; to show it in actual use to his customers, and others who might desire to see it; and also to allow other cooks to be present, and assist when meat was roasted in it, in order to their being convinced of its utility, and taught how to manage it. I likewise prevailed on him to engage an intelligent bricklayer in his service who would submit to be taught to set roasters properly, and who would follow without deviation the directions he should receive. All these arrangements were carried into execution in the beginning of the year 1799; and since that time Mr. Summers has sold and put up no less than 260 roasters, all of which have been found to answer perfectly well; and, although he employs a great many hands in the manufacture of this new article, he is not able to satisfy all the demands of his numerous customers.

Many of these roasters have been put up in the houses of persons of the highest rank and distinction; others in the kitchens of artificers and tradesmen; and others again in schools, taverns, and other houses of public resort; and in all these different situations the use of them has been found to be economical, and advantageous in all respects.

Several other tradesmen in London have also been engaged in the manufacture of roasters. Mr. Hopkins, of Greek Street, Soho, ironmonger to the king, made that which is at the Foundling Hospital, likewise that which was put up in the house formerly occupied by the Board of Agriculture; and he informs me that he has sold above 200 others, which have been put up in the kitchens of various hospitals and private families in the capital and in different parts of the country.

Messrs. Moffat & Co., of Great Queen Street, Lincoln's-Inn Fields, and Mr. Feetham, of Oxford Street, as also Mr. Gregory, Mr. Spotswood, Mr. Hanan, and Mr. Briadwood, in Edinburgh, have engaged in the manufacture of them. Other tradesmen, no doubt, with whose names I am not acquainted, have manufactured them; and as there is no difficulty whatever in their construction, and as all persons are at full liberty to manufacture and sell them, I hope soon to see these roasters become a common article of trade.

I have done all that was in my power to improve and to bring them forward into notice; and all my wishes respecting them will be accomplished if they should be found to be useful, and if the public is furnished with them at reasonable prices.

Several roasters, constructed by different workmen, may be seen, some of them set in brick-work, and others not, at the repository of the Royal Institution.

I have delayed thus long to publish a description of this contrivance, in order that its usefulness might previously be established by experience; and also that I might be able, with the description, to give notice to the public where the thing described might be seen. I was likewise desirous of being able at the same time to point out several places where the article might be had.

These objects having been fully accomplished, I shall now proceed by giving

*An Account of the Roaster, and of the Principles on which it is constructed.*

When I first set about to contrive this machine, meditating on the nature of the mechanical and chemical operations that take place in the culinary process in question, it appeared to me that there could not possibly be any thing more necessary to the roasting of meat than heat in certain degrees of intensity, accompanied by certain degrees of dryness; and I thought if matters could be so arranged, by means of simple mechanical contrivances, that the cook should be enabled not only to regulate the degrees of heat at pleasure, but also to combine any given degree of heat with any degree of moisture or of dryness required, this would unquestionably put it in his power to perform every process of roasting in the highest possible perfection.

The means I used for attaining these ends will appear by the following description of the machinery I caused to be constructed for that purpose.

The most essential part of this machinery, which I shall call the *body* of the roaster (see Fig. 14), is a

Fig. 14.

hollow cylinder of sheet iron (which, for a roaster of a moderate size, may be made about 18 inches in diameter and 24 inches long), closed at one end, and set in a horizontal position in a mass of brick-work, in such a manner that the flame of a small fire, which is made in a closed fire-place directly under it, may play all round it, and heat it equally and expeditiously.   The open end of this cylinder, which should be even with the front of the brick-work in which it is set, is closed either with a double door of sheet iron, or with a single door of sheet iron covered on the outside with a panel of wood; and in the cylinder there is a horizontal shelf, made of a flat plate of sheet iron, which is supported on ledges riveted to the inside of the cylinder, on each side of it.   This shelf is situated about three inches below the centre or level of the axis of the body of the roaster, and it serves as a support for a dripping-pan, in which, or rather over which, the meat to be roasted is placed.

This dripping-pan, which is made of sheet iron, is about 2 inches deep, 16 inches wide above, 15¾ inches in width below, and 22 inches long; and it is placed on four short feet, or, what is better, on two long sliders, bent upwards at their two extremities, and fastened to the ends of the dripping-pan, forming, together with the dripping-pan, a kind of sledge; the bottom of the dripping-pan being raised by these means about an inch above the horizontal shelf on which it is supported.

In order that the dripping-pan on being pushed into or drawn out of the roaster may be made to preserve its direction, two straight grooves are made in the shelf on which it is supported, which, receiving the sliders of the dripping-pan, prevent it from slipping about from side to side, and striking against the sides of the roaster. The front ends of these grooves are seen in Fig. 14, as are also the front ends of the sliders of the dripping-pan, and one of its handles.

In the dripping-pan, a gridiron (seen in Fig. 14) is placed, the horizontal bars of which are on a level with the sides or brim of the dripping-pan, and on this gridiron the meat to be roasted is laid; care being taken that there be always a sufficient quantity of water in the dripping-pan to cover the whole of its bottom to the height of at least half or three quarters of an inch.

This water is essential to the success of the process of roasting: it is designed for receiving the drippings from the meat, and preventing their falling on the heated bottom of the dripping-pan, where they would be evaporated, and their oily parts burned or volatilized, filling the roaster with ill-scented vapours, which would spoil the meat by giving it a disagreeable taste and smell.

It was with a view more effectually to defend the bottom of the dripping-pan from the fire, and prevent as much as possible the evaporation of the water it contains, that the dripping-pan was raised on feet or sliders, instead of being merely set down on its bottom on the shelf which supports it in the roaster.

A late improvement has been made in the arrangement of the dripping-pan, by an ingenious workman at Norwich, Mr. Frost, who has been employed in putting up roasters in that part of the country; an invention which I think will, in many cases, if not in all, be found very useful. Having put a certain quantity of water into the principal dripping-pan, which is constructed of sheet iron, he places a second, shallower, made of tin, and standing on four short feet, into the first, and then places the gridiron which is to support the meat in this second dripping-pan. As the water in the first keeps the second cool, there is no necessity for putting water into this; and the drippings of the meat may, without danger, be suffered to fall into it, and to remain there unmixed with water. When Yorkshire puddings or potatoes are cooked under roasting meat, this arrangement will be found very convenient.

In constructing the dripping-pans, and fitting them to each other, care must be taken that the second do not touch the first, except by the ends of its feet; and especially that the bottom of the second (which may be made dishing) do not touch the bottom of the first. The lengths and widths of the two dripping-pans above, or at their brims, may be equal, and the brim of the second may stand about half an inch above the level of the brim of the first. The horizontal level of the upper surface of the gridiron should not be lower than the level of the brim of the second dripping-pan; and the meat should be so placed on the gridiron that the drippings from it cannot fail to fall into the dripping-pan, and never upon the hot bottom or sides of the roaster.

To carry off the steam which arises from the water in the dripping-pan, and that which escapes from the meat in roasting, there is a steam-tube belonging to the roaster, which is situated at the upper part of the roaster, commonly a little on one side and near the front of it, to which tube there is a damper, which is so contrived as to be easily regulated without opening the door of the roaster. This steam-tube is distinctly seen in Fig. 14; and the end of the handle by which its damper is moved may be seen in Fig. 15.

The heat of the roaster is regulated at pleasure, and to the greatest nicety, by means of the register in the ash-pit door of its fire-place (represented in Fig. 15) and by the damper in the canal, by which the smoke goes off into the chimney, which damper is not represented in any of the figures.

The *dryness* in the roaster is regulated by the damper of the steam-tube, and also by means of a very essential part of the apparatus — the *blowpipes* — which still remain to be described. They are distinctly represented in the Figs. 14, 15, and 16.

Fig. 15.

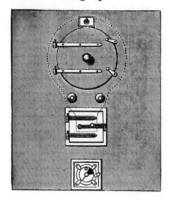

These blowpipes, which lie immediately under the roaster, are two tubes of iron, about $2\frac{1}{2}$ inches in diameter and 23 inches long, or about 1 inch shorter than the roaster; which tubes, by means of elbows at their farther ends, are firmly fixed to the bottom of the roaster, and communicate with the inside of it. The hither ends of these tubes come through the brick-work, and are seen in front of the roaster, being even with its face.

These blowpipes have stoppers, by which they are accurately closed; but when the meat is to be *browned* these stoppers are removed, or drawn out a little, and the damper in the steam-tube of the roaster being at the same time opened a strong current of hot air presses in through the tubes into the roaster, and through the roaster into and through the steam-tube, carrying and driving away all the moist air and vapour out of the roaster.

Fig. 16.

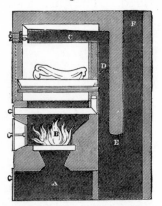

As these blowpipes are situated immediately below the roaster and just over the fire, and are surrounded on every side by the flame of the burning fuel (see Fig. 16), they are much exposed to the heat; and when the fire is made to burn briskly, which should always be done when the meat is to be browned, they will be heated red-hot, consequently the air which passes through them into the roaster will be much heated; and this *hot wind* which blows over the meat will suddenly heat and dry its surface in every part, and give it that appearance and taste which are peculiar to meat that is well roasted.

When these roasters were first proposed, and before their merit was established, many doubts were entertained respecting the taste of the food prepared in them.

As the meat was shut up in a confined space, which has much the appearance of an oven, it was natural enough to suspect that it would be rather *baked* than *roasted;* but all those who have tried the experiment have found that this is by no means the case. The meat is *roasted,* and not *baked;* and, however bold the assertion may appear, I will venture to affirm that meat of every kind, without any exception, roasted in a roaster, is *better tasted, higher flavoured, and much more juicy and delicate* than when roasted on a spit before an open fire.

I should not have dared to have published this opinion four years ago; but I can with safety do it now, for I can appeal for a confirmation of the fact to the results of a number of decisive experiments lately made in this metropolis, and by the most competent judges.

Among many others who, during the last year, have caused roasters to be put up in their kitchens, I could mention one person in particular, a nobleman, distinguished as much by his ingenuity and indefatigable zeal in promoting useful improvements as by his urbanity and his knowledge in the art of refined cookery, who had two roasters put in his house in town, and who informs me that he has frequently invited company to dine with him since his roasters have been in use, and that the dishes prepared in them have never failed to meet with marked approbation.

In enumerating the excellences of this new implement of cookery, there is one of indisputable importance, which ought not to be omitted. When meat is roasted in this machine, its quantity, determined by weight, is considerably greater than if it were roasted upon a spit before a fire. To ascertain this fact, two legs of mutton taken from the same carcass, and made perfectly equal in weight before they were cooked, were roasted on the same day, the one in a roaster, the other on a spit before the fire; and, to prevent all deception, the persons employed in roasting them were not informed of the principal design of the experiment. When these pieces of roasted meat came from the fire they were carefully weighed; when it appeared that the piece which had been roasted in the roaster was heavier than the other by a difference which was equal to six per cent, or six pounds in a hundred. But this even is not all; nor is it indeed the most important result of the experiment. These two legs of mutton were brought upon table at the same time, and a large and perfectly unprejudiced company was assembled to eat them. They were both declared to be very good; but a decided preference was unanimously given to that which had been roasted in the roaster, it was much more juicy, and was thought better tasted. They were both fairly eaten up, nothing remaining of either of them that was eatable. Their fragments, which had been carefully preserved, being now collected and placed in their separate dishes, it was *a comparison of these fragments* which afforded the most striking proof of the relative merit of these two methods of roasting meat, in respect to the economy of food. Of the leg of mutton which had been roasted in the roaster, hardly any thing visible remained except the bare bone; while a considerable heap was formed of scraps not eatable which remained of that roasted on a spit.

I believe I may venture to say that the result of this experiment is deserving of the most serious attention, especially in this country, where so much roasted meat is eaten, and where the economy of food is every day growing to be more and more an object of public concern.

I could mention several other experiments similar to that just described, which have been made, and with similar results; but it would be superfluous to bring many examples to ascertain a fact which is so well established by one.

There is one peculiarity more respecting meat roasted in a roaster, which I must mention; that is, the uncommon delicacy of the taste of the fat of the meat so roasted, especially when it has been done by a very slow fire. When good mutton is roasted in this manner, its fat is exquisitely sweet and well tasted, and when eaten with currant jelly can hardly be distinguished from the fat of the very best venison. The fat parts of other kinds of meat are also uncommonly delicate when prepared in this manner; and there is reason to think that they are much less unwholesome than when they are roasted before an open fire.

The heat which is generated by the rays which proceed from burning fuel is frequently most intense; and hence it is that the surface of a piece of meat that is roasted on a spit is often quite burned, and rendered not only hard and ill-tasted, but very unwholesome. The fat of venison is not thought to be unwholesome; but, in roasting venison, care is taken, by covering it, to prevent the rays from the fire from burning it. In the roasting machine, the bad effects of these direct rays are always prevented by the sides of the roaster, which intercepts them, and protects the surface of the meat from the excessive violence of their action; and even when, at the end of the process of roasting, the intensity of the heat in the roaster is so far increased as to brown the surface of the meat, yet this heat being communicated through the medium of a heated fluid (air) is much more moderate and uniform and certain in its effects, than direct rays which proceed from burning fuel, or from bodies heated to a state of incandescence.

### *Directions for setting Roasters.*

There are two points to which attention must be paid by bricklayers in setting these roasters, otherwise they will not be found to answer. Their fire-places must be made extremely small; and provision must be made for cleaning out their flues from time to time when they become obstructed with soot.

When I first introduced these roasters into this country five years ago, I was not fully aware of the irresistible propensity to make too great fires on all occasions, which those people have who inhabit kitchens; but sad experience has since taught me that nothing short of rendering it absolutely impossible to destroy my roasters by fire will prevent their being so destroyed. The knowledge of this fact has put me on my guard, and I now take effectual measures for preventing this evil. I cause the fire-places of roasters to be made very small, and direct them to be situated at a considerable distance below the bottom of the roaster.

For a roaster which is 18 inches wide, and 24 inches long, the fire-place should not be more than 7 inches wide and 9 inches long; and the side walls of the fire-place should be quite vertical to the height of 6 or 7 inches. Small as this fire-place may appear to be, it will contain quite coals enough to heat the roaster, and many more than will be found necessary for keeping it hot when heated. The fact is that the quantity of fuel required to roast meat in this way is almost incredibly small. By experiments made with great care at the Foundling Hospital, it appeared to be only about one sixteenth part of the quantity which would be required to roast the same quantity of meat in the common way before an open fire. But it is not merely to save fuel that I recommend the fire-places to be made very small: it is to prevent the roasters from being wantonly destroyed, the meat spoiled, and a useful invention discredited.

With regard to the provision which ought to be made, in the setting of a roaster, for occasionally cleaning out its flues, this must be done by leaving proper openings (about 4 or 5 inches square, for instance) in the brick-work, to introduce a brush, like a bottle-brush, with a long handle; which openings may be closed with stoppers or fit pieces of brick or of stone, and the joinings made good with a little moist clay. To render these stoppers more conspicuous, they may each be furnished with a small iron ring or knob, which will likewise be useful as a handle in removing them and replacing them.

In Figs. 15 and 16, a simple contrivance may be seen represented, by means of which the soot which is apt to collect about the top of a roaster may be removed with very little trouble as often as it shall be found necessary, without injuring the brick-work or deranging any part of the machinery. By means of an oblong square frame, constructed of sheet iron, and fastened to the top of the roaster by rivets, a door-way is opened into the void space left for the flame and smoke between the outside of the roaster and the hollow arch or vault in which it is placed; and by introducing a brush with a flexible handle through this door-way, the soot adhering to the outside of the top of the roaster, and to the surface of the brick-work surrounding it, may be detached and made to fall back into the fire-place, from whence it may be removed with a shovel. The sides of the roaster may be cleaned by introducing a brush through the door-way of the fire-place.

The door-way at the top of the roaster may be closed either by a stopper made of sheet iron, or by a fit piece of stone or brick, furnished with a ring or knob to serve as a handle to it.

If cokes be burned under these roasters, instead of coal (which, as they will not be more expensive fuel, and as they burn longer, and give a more equal heat, I would strongly recommend), the flues will seldom if ever require to be cleaned out. I burn nothing but coke and a few pieces of wood in the closed fire-places of my own kitchen; and for my open chimney fires I use a mixture of coke and coals, which makes a very pleasant fire, and is, I believe, less expensive than coals. It appears to me that there is no subject which offers so promising a field for experimental investigation, and where useful improvements would be so likely to be made, as in the *combination and preparation of fuel.* But to return from this digression.

In constructing the fire-place of a roaster (and all other closed fire-places) care must be taken to place the iron bars on which the fuel burns at a considerable distance from the door of the fire-place; otherwise, this door being near the fire, its handle will become very hot, and it will burn the hand of a person that takes hold of it. I have more than once seen roasters and ovens condemned, disgraced, and totally neglected, merely from an accident of this kind. And yet how easy would it have been to have corrected this fault! If the door of the fire-place is found to become too hot, send for the bricklayer, and let him put the fire-place farther backward.

There should always be a passage or throat, of a certain length, between the mouth or door of a closed fire-place and the fire-place properly so called, or the cavity occupied by the burning fuel. Where fire-places are of large dimensions, it is very useful (as indeed it is customary) to keep this throat constantly filled and choked up with coal. This coal, which, as there is no supply of air in the passage, does not burn, serves to defend the fire-place door from the heat of the fire. It serves another useful purpose : it gets well warmed, and even heated very hot, before it is pushed forward into the fire-place, which disposes it to take fire instantaneously, and without cooling the fire-place and depressing the fire when it is introduced. If any part of it takes fire while it occupies the throat or passage of the fire-place, it is that part only which is in immediate contact with the burning fuel, and what is so burned is consumed under the most advantageous circumstances; for the thick vapour which rises from this coal, as it grows very hot, and which under other less favourable circumstances would not fail to go off in smoke, takes fire in passing over the burning fuel, and burns with a clear bright flame. I have had frequent opportunities of verifying this interesting fact; and I mention it now, in order, if possible, to fix the attention of those who have the management of large fires, to an object which perhaps is of greater importance than they are aware of.

When good reasons can be assigned for the advantages which result from any common practice, this not only tends to satisfy the mind, and make people careful, cheerful, and attentive in the prosecution of their business, but it has also a very salutary influence, by preventing those perpetual variations and idle attempts at improvement, *undirected by science*, which are the consequence of the inconstancy, curiosity, and restlessness of man.

Discoveries are always accidental; and the great use of science is by investigating the nature of the effects produced by any process or contrivance, and of the causes by which they are brought about, to explain the operation and determine the precise value of every new invention. This fixes as it were the *latitude* and *longitude* of each discovery, and enables us to place it in that part of the map of human knowledge which it ought to occupy. It likewise enables us to use it in taking *bearings* and *distances*, and in shaping our course when we go in search of new discoveries. But I am again straying very far from my humble subject.

In constructing closed fire-places for roasters, boilers, ovens, etc., for kitchens, I have found it to be a good general rule to make the distance between the fire-place door and the hither end of the bars of the grate just equal to the width of the fire-place, measured just above the bars. In fire-places of a moderate size, where double doors are used, it will suffice if the distance from the hinder side of the inner door to the hither end of the bars be made equal to the width of a brick, or $4\frac{1}{2}$ inches; but, if the door be not double, it is necessary that the length of the passage from the door into the place occupied by the burning fuel should be at least 6 or 7 inches.

In setting the iron frame of the door of a closed fire-place, care should be taken to mask the metal by setting the bricks before it in such a manner that no part of the frame *may be seen* (if I may use that expression) by the fire. This precaution should be used in constructing fire-places of all sizes, otherwise the frame of the fire-place door will be heated very hot by the rays from the burning fuel, especially when the fire-place is large, and its form will soon be destroyed by the frequent expansion and contraction of the metal. The consequences of this change of form will be the loosening of the frame in the brick-work, and the admission of air into the fire-place over the fire between the sides of the frame and the brick-work, and likewise between the frame and its door, which will no longer fit each other.

The expense of keeping large fire-places in repair is very considerable, as I have learned from some of the London brewers. More than nine tenths of that expense might easily be saved by constructing the machinery more scientifically, and using it with care.

Fig. 15,.page 157, is a front view; and Fig. 16, page 158, represents a vertical section of a roaster, set in brick-work. The hollow spaces represented in Fig. 16 are expressed by strong vertical lines; namely, the ash-pit, A; the fire-place, B; the space between the outside of the roaster and the arch of brick-work which surrounds it, C; the broad canal at the farther end of the roaster, by which the smoke descends, D; and the place E, where it turns, in order to pass upwards into the chimney by the perpendicular canal, F. The brick-work is expressed by fainter lines drawn in the same direction.

The farther end of the roaster must be so fixed in the brick-work that no part of the smoke can find its way from the fire-place, B, directly into the canal, D, otherwise it will not pass up by the sides of the roaster to the top of it. At the top of the roaster, at its farther end, an opening must of course be left for the smoke to pass into the descending canal, D.

As I have already mentioned the necessity of causing the smoke which is used for heating an iron oven or a roaster *to descend* before it is permitted to pass off into the chimney, I shall insist no farther on that important point in this place. It may, however, be useful to observe that, if the place where a roaster is set is not deep enough to allow of the descending canal, D, and the canal, F, by which the smoke ascends and passes into the chimney, to be situated at the farther end of the roaster, both these canals may, without the smallest inconvenience, be placed on one side of the roaster; indeed, as houses are now built, it will commonly be most convenient to place them on one side, and not at the end of the roaster. When this is done, the smoke must be permitted to pass up behind the farther end of the roaster, as well as by the sides of it.

By taking away a large flat stone, or a twelve-inch tile, placed edgeways, a passage from A to E may be opened occasionally, in order to clean out the canals, D and F, and remove the soot. These passages may be cleaned out either from above or from below, by means of a brush with a long flexible handle.

The steam-tube (which is seen in this figure) must open into a separate canal (not expressed in the figure), which must be constructed for the sole purpose of carrying off the steam into the chimney or into the open air. If this steam-tube were to open into either of the cavities or canals, C, D, E, or F, in which the smoke from the fire which heats the roaster circulates, this smoke might, on some occasions, be driven back into the roaster, which could not fail to give a bad taste to the meat. The steam-tube must be laid on a descent, otherwise the water generated in it, in consequence of the condensation of the steam, might run back into the roaster.

Some care will be necessary in forming the vault which is to cover the roaster above. Its form should be regular, in order that it may be everywhere at the same distance from the roaster; and its concave surface should be as even and smooth as possible, in order that there may be the fewer cavities for the lodgement of soot. The distance between the outside of the roaster and the concave surface of this vault may be about 2 inches; and the same distance may be preserved below, between the brick-work and the sides of the roaster. In the Fig. 15 the outline of the fire-place and of the cavity in which the roaster is set is indicated by a dotted line.

### Directions for the Management of a Roaster.

Care must be taken to keep the roaster very clean, and, above all, to prevent the meat from touching the sides of it, and the gravy from being spilt on its bottom. If by any means it becomes greasy in any part that is exposed to the action of the fire, as the metal becomes hot this grease will be evaporated, as has already been observed, and will fill the roaster with the most offensive vapour. When grease spots appear, the inside of the roaster must be washed, first with soap and water to take away the grease, and then with pure water to take away the soap, and it must then be wiped with a cloth till it be quite dry.

The fire must be moderate, and time must be allowed for the meat to be roasted *by the most gentle heat.* About one third more time should in general be employed in roasting meat in a roaster, than would be necessary to roast it in the usual way, on a spit before a fire.

The blowpipes should be kept constantly closed from the time the meat goes into the roaster till within 12 or 15 minutes of its being sufficiently done to be sent to the table; that is to say, till it is fit *to be browned.*

The meat is browned in the following manner: the fire is made to burn bright and clear for a few minutes, till the blowpipes begin to be red-hot (which may be seen by withdrawing their stoppers for a moment, and looking into them), when the damper of the steam-tube of the roaster being opened, and the stoppers of the blowpipes drawn out, a certain quantity of air is permitted to pass through the heated blowpipes into and through the roaster.

I say a certain quantity of air is allowed to pass through the blowpipes into the roaster. If the steam-tube and the blowpipes were set wide open, it is very possible that too much might be admitted, and that the inside of the roaster and its contents might be cooled by it, instead of being raised to a higher temperature. As the velocity with which the cold air of the atmosphere will rush into and through the blow-pipes of a roaster will depend on a variety of circumstances, and may be very different even in roasters of the same size and construction, no general rules can be given in browning the meat for the regulation of the stoppers of the blowpipes, and of the damper in the steam-tube: these must depend on what may be called *the trim of the roaster*, which will soon be discovered by the cook.

There is an infallible rule for the regulation of the damper of the steam-tube, *during the time the meat is roasting by a gentle heat.* It must then be kept just so much opened that the steam which arises from the meat, and from the evaporation of the water in the dripping-pan, may not be seen coming out of the roaster through the crevices of its door; for, if it be more opened, the cold air of the atmosphere will rush into the roaster through those crevices, and by partially cooling it will derange the process that is going on; and, if it be less opened, the room will be filled with steam.

In brightening the fire, preparatory to the browning of the meat, the register in the ash-pit door, and the damper in the canal by which the smoke passes off into the chimney, should both be opened; and it may be useful to stir up the fire with a poker, but this would be a very improper time for throwing a quantity of fresh coals into the fire-place, for that would cool the fire-place, and damp the fire for a considerable time. By far the best method of brightening the fire for this purpose would be to throw a small fagot into the fire, or a little bundle of dry wood of any kind, split into small pieces about six or seven inches in length. This would afford a clear bright flame, which would heat the blowpipes quickly, and without injuring them. Indeed, wood ought always to be used for heating roasters, in preference to coal, where it can be had; and the quantity of it required is so extremely small, that the difference in the expense would be very trifling, even here in London, where the price of fire-wood is so high. And if the durability of the machinery be taken into the account, which is but just, I am confident that, for heating roasters and ovens constructed of sheet iron, coals would turn out to be dearer fuel than wood.

I have already insisted so much on the necessity of keeping a quantity of water under meat that is roasting, in order to prevent the drippings from the meat from falling on any very hot metal, that I shall not now enlarge farther on the subject, except by saying once more that it is a circumstance to which it is indispensably necessary to pay attention.

When meat is roasted by a very moderate heat, it will seldom or never require being either turned or basted; but, when the heat in the roaster is more intense, it will be found useful both to turn it and to baste it three or four times during the process. The reason of this difference in the manner of proceeding will be evident to those who consider the matter with attention.

When roasters are constructed of large dimensions, several kinds of meat may be roasted in them at the same time. If care be taken to preserve their drippings separate, which may easily be done by placing under each a separate dish or dripping-pan, standing in water contained in a larger dripping-pan, there will be no mixture of tastes; and, what no doubt will appear still more extraordinary, a whole dinner, consisting of various dishes, — roasted, stewed, baked, and boiled, — may be prepared at the same time in the same roaster, without any mixture whatever of tastes. A respectable friend of mine who first made the experiment, and who has since repeated it several times, has assured me of this curious fact. It may, perhaps, in time turn out to be an important discovery. A simple and economical contrivance, by means of which all the different processes of cookery could be carried on at the same time and by one small fire, would, no doubt, be a valuable acquisition.

It is very certain that roasters will either bake or roast separately in the highest possible perfection; and it is not improbable that, with certain precautions in the management of them, they may be made to perform those two processes at the same time, in such a manner as to give general satisfaction. When roasters are designed for roasting and baking at the same time, they should be made sufficiently large to admit of a shelf above the meat, on which the things to be baked should be placed. I am told that above half the roasters lately put up in London are so constructed, and that they are frequently made to roast and bake at the same time. I shall take another opportunity of enlarging on the utility of this contrivance.

There is a precaution to be taken in opening the door of a roaster, when meat is roasting in it, which ought never to be neglected; that is, to open the steam-tube and both the blowpipes, for about a quarter of a minute, or while a person can count fifteen or twenty, before the door of the roaster be thrown open. This will drive away the steam and vapour out of the roaster, which otherwise would not fail to come into the room as often as the door of the roaster is opened.

As it will frequently happen that the meat will be done before it will be time to send it up to table, when this is the case, it may either be taken out of the roaster and put into a hot closet, which may very conveniently be situated immediately over the roaster, or it may remain in the roaster till it is wanted. If this last-mentioned method of keeping it warm be adopted, the following precautions will be necessary for cooling the roaster, otherwise the process of roasting will still go on, and the meat, instead of being merely kept warm, will be over done. The register in the ash-pit door should be closed; the fire-place door and the damper in the chimney should be set wide open; the fire should either be taken out of the fire-place or it should be covered with cold ashes; and, lastly, the damper in the steam-tube and both the blowpipes should be opened. By these means the heat will very soon be driven away up the chimney, and, as soon as it is so far moderated as to be no longer dangerous, the blow-pipes and the damper in the steam-tube may be nearly closed; and if there should be danger of the cooling being carried too far, the fire-place door may be shut. By these means the heat of the roaster and of the brick-work which surrounds it may be moderated and regulated at pleasure; and meat already roasted may be kept warm, for almost any length of time, without any danger of its being spoiled.

### *Miscellaneous Observations respecting Roasters and Ovens.*

I shall, no doubt, be criticised by many for dwelling so long on a subject which to them will appear low, vulgar, and trifling; but I must not be deterred by fastidious criticisms from doing all I can do to succeed in what I have undertaken. Were I to treat my subject superficially, my writings would be of no use to anybody, and my labour would be lost; but, by investigating it thoroughly, I may perhaps engage others to pay that attention to it which, from its importance to society, it certainly deserves. If improvements in articles of elegant luxury, which not one person in ten thousand is rich enough to purchase, are considered as matters of public concern, how much more interesting to a benevolent mind must those improvements be which contribute to the comfort and convenience of every class of society, rich and poor.

But the subject now under consideration is very far from being uninteresting, even if we consider it merely as it is connected with science, without any immediate view to its utility; for in it are involved several of the most abstruse questions relative to the doctrine of heat.

Many have objected to the roaster, on the supposition that meat cooked in it must necessarily partake more of the nature of baked meat than of roasted meat. The general appearance of the machinery is certainly calculated to give rise to that idea, and when it is known that all kinds of baking may be performed in great perfection in the roaster, that circumstance no doubt tended very much to confirm the suspicion; but, when we examine the matter attentively, I think we shall find that this objection is not well founded.

When any thing is baked in an oven (on the common construction), the heat is gradually diminishing during the whole time the process is going on. In the roaster, the heat is regulated at pleasure, and can be suddenly increased towards the end of the process; by which means the distinguishing and most delicate operation, *the browning of the surface* of the meat, can be effected in a few minutes, which prevents the drying up of the meat and the loss of its best juices.

In an oven, the exhalations being confined, the meat seldom fails to acquire a peculiar and very disagreeable smell and taste, which, no doubt, is occasioned solely by those confined vapours. The steam-tube of a roaster being always set open, when in browning the meat the heat is sufficiently raised to evaporate the oily particles at its surface, the noxious vapours unavoidably generated in that process are immediately driven away out of the roaster by the current of hot and pure air from the blow-pipes. This leaves the meat perfectly free both from the taste and the smell peculiar to baked meat.

Some have objected to roasters, on an idea that, as the water which is placed under the meat is (in part at least) evaporated during the process, this must make the meat sodden, or give it the appearance and taste of meat boiled in steam; but this objection has no better foundation than that we have just examined. As steam is much lighter than air, that generated from the water in the dripping-pan will immediately rise up to the top of the roaster, and pass off by the steam-tube, and the meat will remain surrounded by air, and not by steam. But were the roaster to be constantly full of steam, to the perfect exclusion of all air, which however is impossible, this would have no tendency whatever to make the meat sodden. It is a curious fact that steam, so far from being a moist fluid, is perfectly dry, as long as it retains its elastic form; and that it is of so drying a nature that it cannot be contained in wooden vessels (however well seasoned they may be) without drying them and making them shrink till they crack and fall to pieces.

Steam is never moist. When it is condensed with cold, it becomes water, which is moisture itself, but the steam in a roaster, which surrounds meat that is roasting, cannot be condensed upon it; for the surface of the meat, being heated by the calorific rays from the top and sides of the roaster, is even hotter than the steam.

If steam were a moist fluid, it would be found very difficult to bake bread, or any thing else, in a common oven.

Meat which is *boiled* or *sodden* in steam is put cold into the containing vessel, and the hot steam which is admitted is instantly condensed on its surface, and the water resulting from this condensation of steam dilutes the juices of the meat and washes them away, leaving the meat tasteless and insipid at its surface; but when meat is put cold into a roaster, the water in the dripping-pan being cold likewise, long before it can acquire heat sufficient to make it boil, the surface of the meat will become too hot for steam to be condensed upon it; and, were it not to be browned at all, it could not possibly taste sodden.

It appears to me that these elucidations are sufficient to remove the two objections which are most commonly made to the roaster by those who are not well acquainted with its mechanism and manner of acting.

In my account of the blowpipes, I have said that the current of air which comes into the roaster through them, when they are opened to brown the meat, "drives away all the moist air and vapour out of the roaster." This I well know is not an accurate account of what really happens; but it may serve, perhaps better than a more scientific explanation, to give the generality of readers distinct ideas of the nature of the effects that are produced by them.   The noxious vapour generated from the oily particles that are evaporated by the strong heat are most certainly driven away precisely in the manner described; and we have just seen how very essential it is that these vapours should not be permitted to remain in the roaster.   And whether the surface of the meat be in fact dried by the immediate contact of a current of hot and dry air, or whether this effect is produced in consequence of an increase of calorific rays from the top and sides of the roaster occasioned by the additional heat communicated to the internal surface of the roaster by this hot wind, the utility of the blowpipes is equally evident in both cases.

# CHAPTER V.

*More particular Descriptions of the several Parts of the Roaster, designed for the Information of Workmen. — Of the Body of the Roaster. — Of the Advantages which result from its peculiar Form. — Of the best Method of proceeding in covering the iron Doors of Roasters and Ovens, with Panels of Wood, for confining the Heat. — Method of constructing double Doors of sheet Iron and of cast Iron. — Of the Blowpipes. — Of the Steam-tube. — Of the Dripping-pan. — Precautions to be used for preventing the too rapid Evaporation of the Water in the Dripping-pan. — Of large Roasters that may be used for roasting and baking at the same Time. — Precautions which*

*become necessary when Roasters are made very large.
— Of various Alterations that may be made in the
Forms of Roasters, and of the Advantages and
Disadvantages of each of them. — Account of some
Attempts to simplify the Construction of Roasters.
— Of a Roasting-oven. — Of the Difference between
a Roasting-oven and a Roaster.*

ALTHOUGH it will be easy for persons acquainted
with the mechanic arts, and accustomed to examine drawings and descriptions of machines, to form a
perfect idea of the invention in question from what has
already been said, yet something more will be necessary
for the instruction of artificers who may be employed
in executing the work, and more especially for such
as may from these descriptions undertake to construct
roasters without ever having seen one. By going into
these details, I shall no doubt find opportunities for
introducing occasional remarks on the uses and management of the various parts of the machinery, which
will tend not a little to illustrate the foregoing descriptions, and enable the reader to form a more precise
and satisfactory opinion respecting the merit of the
contrivance.

## Of the Body of the Roaster.

Although I have directed the body of the roaster to
be made cylindrical, it may, without any considerable
inconvenience, be constructed of other forms. The
reasons why I preferred the cylindrical form to all
others were because I was told by workmen that it
was the form of easiest construction; and because I

knew it to be the form best adapted for strength and durability.

There is another reason, which I did not dare to communicate to the workmen (iron-plate workers) whom I was obliged to employ, in order to introduce this contrivance into common use in this country: when roasters are of this form, it will be easy to make them of cast iron, which will render the article not only cheaper to the purchaser, but also much more durable, and better on many accounts.

As there is a certain proportion of sulphur in the coal commonly used in this country, I was always perfectly aware of the consequences of burning it under roasters constructed of sheet iron. I knew that the sulphurous vapour from such fuel would be much more injurious to the roaster, and especially to its blow-pipes (which are much exposed) than the clear flame of a wood fire; but I trusted to the remedy, which I knew might easily be provided for this defect. I thought that cast iron, which is much less liable to be injured by a coal fire than wrought iron, would soon be substituted in lieu of it, first for the blowpipes, and then for the body of the roaster. In this expectation I have not been disappointed, for the blowpipes of roasters are now commonly made of cast iron by the London workmen; and, where sea-coal is used as fuel, they never should be made of any other material.

The first roasters I caused to be made had all flat bottoms, and their sides were vertical, and their tops were arched over in the form of a trunk; but several inconveniences were found to result from this shape. Their bottoms were too much exposed to the heat, and this excessive heat in that part heated the bottom of the dripping-pan too much, and caused the water in it to be soon evaporated; it likewise caused them to warp, and sometimes prevented their doors from closing them with that precision which is necessary.

If the hot air in a roaster be permitted to escape by the crevices of its door, or, what is still worse and more likely to happen, if cold air be permitted to enter the roaster by those openings, it is quite impossible that the process of roasting can go on well.

As cold air will always tend to press into the body of the roaster by every passage that is left open, whenever, the roaster being hot, the damper of its steam-tube is open, — this shows how necessary it is, in roasting meat, not to leave that damper open at any time when it ought to be kept closed.

As iron doors for confining heat are very liable to be warped by the expansion of the metal, they should never be made to shut into grooves, but they should be made to close tight by causing the flat surface of the inside of the door to lie against and touch in all parts the front edge of the door frame, which front edge must of course be made to be perfectly level, and as smooth as possible.

When the body of the roaster is made cylindrical, it will be easier to make the front of it, against which its door closes, level, than if it were of any other form; and when the door is circular, by making it a little dishing, it will not be liable to be warped, especially when it is made double.

If the front end of the cylinder of sheet iron which forms the body of the roaster be turned outwards over a very stout iron wire (about one third of an inch in diameter, for instance), this will strengthen the roaster very much, and will render it easier to make the end of the roaster level to receive the flat surface of its door : it can most easily be made level by placing the cylinder in a vertical or upright position, with its open end downwards, on a flat anvil, and hammering the wire above mentioned till its front edge, which reposes on the anvil, is quite level.

In order that the door of the roaster may close well, its hinges should be made to project outwards two or three inches beyond the sides of the roaster; and it should be fastened not by a common latch, but by two turn-buckles, situated just opposite to the two hinges. The distance at which the two hinges (and consequently the two turn-buckles) should be placed from each other should be equal to half the diameter of the roaster.

The hooks for the hinges, and also the support for the turn-buckles, should be situated at the projecting ends of strong iron straps, fastened at one of their ends to the outside of the roaster, by means of riveting-nails. The manner in which these turn-buckles are constructed, and the manner in which they are fastened to the roaster, may be seen by examining Fig. 17, where they are represented on a large scale.

The first roasters that were made were furnished with two separate doors, the one placed about four inches within the body of the roaster, the other even with its front. As the inside door had no hinges, but, like a common oven door, was taken quite away when the roaster was opened, there was some trouble in the management of it; and it was found that the cooks, to avoid that trouble, frequently threw it away, and used the roaster without it. This contrivance of the cooks to save trouble came very near to discredit the roasters altogether, and to put a final stop to their introduction in this country. The circumstance upon which the principal merit of the roaster depends, and on which the excellence of the food cooked in it depends entirely, is the *equality of the heat.* When the heat is equal on every side, it may be more moderate than when it is unequal; and the more moderate and equal the heat is by which meat can be properly roasted, the better tasted

Fig. 17.

and more wholesome will it be. Now it is quite impossible to keep up an equal heat in a roaster which is closed only by a single door of sheet iron; for so much heat will pass off through such a thin metallic door, and be carried away by the cold air of the atmosphere which is lying against the outside of it, that the degrees of heat in different parts of the roaster must necessarily be very different; and the consequence of this inequality will be either that the meat will not be sufficiently done in some parts, or that the heat must be so much increased as to prevent its being well done in any part.

In order to induce persons to be careful in the management of machinery of any kind which is new to them, it is necessary to point out the bad consequences which will result from such neglects and inattentions as they are most liable to fall into in the use of it; for, however particular instructions may be, strict attention to them cannot be expected from those who are not aware of the bad effects that may result from what may appear to them very trifling deviations or neglects.

Those who make roasters must take the greatest care to construct them in such a manner that they may be accurately closed, and that the heat may not be able to make its way through their doors; and those who use them must be careful to manage them properly.

There are two ways in which the door of a roaster may be constructed, so as to confine the heat perfectly well, without giving any additional trouble to the cook in the management of it. It may be made of a single sheet of iron, and covered on the outside with a panel of wood; or it may be constructed of two sheets of iron, placed parallel to each other at the distance of about an inch, and so fastened together that the air between them may be confined.

When a docr of single sheet iron is made to confine the heat by means of an outside covering of wood, care must be taken to make such outside wooden covering in the form of a panel, otherwise it will not answer. If a board be used instead of a framed panel, it will

most certainly warp with the heat, and will either detach itself from the iron door to which it is fastened,
or will cause the door to bend and prevent its closing
the roaster with sufficient accuracy.   I have seen several attempts made to use boards instead of panels,
in covering the outsides of the iron doors of roasters
and iron ovens; but they were all unsuccessful.   It is
quite impossible that they ever should answer, as will
be evident to those who will take the trouble to consider the matter with attention.

As doors of sheet iron, covered with wood on the
outside, when they are properly constructed, are admirably calculated for confining heat, I think it worth
while to give a detailed account of the precautions that
are necessary in the construction of them.

*Of the best Method of covering the iron Doors of*
*Roasters and Ovens, etc., with wooden Panels, for*
*confining the Heat.*

The object principally to be attended to in this business is to contrive matters so that the shrinking and
swelling of the wood by alternate heat and moisture
shall have no tendency either to detach the wood from
the iron door, or to change its form, or to cause openings in the wood by which the air confined between the
wood and the iron can make its escape.

The manner in which this may, in all cases, be done,
will be evident from an examination of the Fig. 18,
which represents a front view of the door of a cylindrical roaster, 18 inches in diameter, covered with a
square wooden panel.

It will be observed that this panel consists of a
square frame tenanted, and fastened together at each

of its four corners with a single pin; and filled up in the middle with a square board or panel, which is confined in its place, by being made to enter into deep grooves or channels, made to receive it, in the insides of the pieces which form the frame. The circular iron door to which this panel is fixed cannot be seen in the figure, being covered and concealed from view by the

Fig. 18.

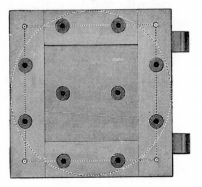

wood, but its size and position are marked out by a dotted circle; and the heads of ten rivets are seen, by which the wooden panel is fastened to the iron door. These rivets are made to hold the wood fast to the iron by means of small circular plates of sheet iron, which are distinctly represented in the figure.*

If the positions of the pins by which the wooden frame is fastened together, and of the rivets which fasten the panel to the iron door, are considered, it will be evident that all bad effects of the shrinking of the wood by the heat are prevented by the proposed

* Instead of these rivets, short wood screws may be used for fastening the wooden panel to the iron door; but care must be taken to place these screws in the same places which are pointed out for the rivets. The heads of the wood screws must of course be on the inside of the iron door.

construction. The four pieces of wood which constitute the frame of the panel (which may be of common deal, and about four inches wide and one inch
thick), being fastened with one pin only at each of
their joinings at the corners, and these pins being situated in the centre of those joinings, if upon the frame,
in the middle of each of the four pieces which compose
it, a square be drawn in such a manner that the corners
of this square may coincide with the centres of the four
pins which hold the frame together, as neither heat nor
dryness makes any considerable alteration in the length
of the fibres of wood, it is evident that the shrinking of
the four pieces which compose this frame cannot alter
the dimensions of this square, or in any way change its
position. If, therefore, care be taken in fastening the
panel to the iron door to place the riveting-nails *in
the lines which form the four sides of this square*, the
shrinking of the wood will occasion no strain on the
iron door, nor have any tendency whatever to change
its form ; and with regard to the centre piece of the
panel, if it be fastened to the iron door by two rivets,
situated *in the direction of the fibres of the wood*, in
a line dividing this piece into two equal parts, its
shrinking will be attended with no kind of inconvenience. Care should, however, be taken to make this
panel enter so deeply into the grooves in its frame
that, when it has shrunk as much as possible, its width
shall not be so much reduced as to cause it to come
quite out of the grooves. This piece may be made
about one third of an inch thick, and the grooves
which receive it may be made of the same width, and
about three quarters of an inch deep.

When wooden covers of this kind are made for iron

doors of large dimensions, they should be divided into a number of compartments, otherwise the centre pieces, or the panels properly so called, being very large, the shrinking of the wood with heat will be apt to make them quit the grooves of their frames, which would open a passage for the cold air to approach the surface of the iron door.

In fastening the wooden panel to its iron door, it will be best that the wood should not come into immediate contact with the iron. Two or three sheets of cartridge paper, placed one upon the other, may be interposed between them; and, to prevent the possibility of this paper taking fire, it may previously be rendered incombustible by soaking it in a strong solution of alum, mixed with a little Armenian bole or common clay. This paper will not only assist very much in confining the heat, but will also effectually prevent the wood from being set on fire by heat communicated through the iron door of the roaster. It is, indeed, highly improbable that the roaster should ever be so intensely heated as to produce this effect; but, as the strangest accidents sometimes do happen, it is always wise to be prepared for the worst that can happen.

As the centre piece of wood, or panel properly so called, which fills up the wooden frame, is only one third of an inch in thickness, while the frame is one inch in thickness, it is evident that, if the face of the frame be made to apply everywhere to the flat surface of the iron door, the centre piece will not touch it. This circumstance will be rather advantageous than otherwise, in confining the heat; but still it will require some attention in fastening the wood to the iron. Each of the two rivets which pass through this centre piece must also be made to pass through a small block of wood, about an inch square for instance, and one third of an inch thick, which will give these rivets a proper bearing, without any strain on the iron door which can tend to alter its form.

When the wood and the iron are firmly riveted together, the superfluous paper may be taken away with a knife.

The hinges of the door, which in the Fig. 18 are seen projecting outwards on the right hand, are to be riveted to the outside surface of the circular iron door; and, in order that they may not prevent the panel from applying properly to the door, they are to be let into the wood. The turn-buckles, by which the door is fastened, must be made to press against the outside or front of the wooden frame.

No inconvenience of any importance will arise from leaving the wooden panel square, while the door itself is circular; but, if it should be thought better, the corners of the panels may be taken off, or the wooden panel may be made circular. This should not, however, be done till after the panel has been fixed to the door. After this has been done, as the rivets will be sufficient to hold the sides of the frame in their places, the cutting off of the corners of the frame will produce no bad consequences.

I have been the more particular in my account of the manner of covering iron doors with wooden panels, for the purpose of confining heat, as this contrivance may be used with great advantage, not only for roasters and ovens, but also for a variety of other purposes; for the covers of large boilers, for instance, for the doors of hot closets, steam closets, etc.

*Of double Doors for Roasters, constructed of two circular Pieces of sheet Iron seamed together.*

No difficulty will be found in the construction of these doors; and though they may not, perhaps, confine the heat quite so perfectly as the doors we have just described, they answer very well; and, when the outside of the door is japanned, they have a very handsome and cleanly appearance.

There are two ways of constructing them, either of which may be adopted: the circular sheet of iron which forms the inside of the door may be flat and the outside sheet dishing, or the outside sheet may be flat and the inside sheet dishing; but, whichever of these methods is adopted, the hinges must be attached to the outside of the door, and care must be taken to make that part of the inside of the door quite flat which lies against the end of the roaster, and closes it. The distance of the inside sheet of iron and the outside sheet is not very essential: it should not, however, be less than one inch in the centre of the door; and these two sheets should not touch each other anywhere, except it be at their circumference, where they are fastened together. In the centre of the outside sheet there should be fixed a knob of iron or of brass, to serve as a handle for opening and shutting the door.

Double doors of this kind might easily be constructed of two circular pieces of cast iron, fastened together by rivets; or of one piece of cast iron, cast dishing, and a flat piece of sheet iron turned over it. When the latter construction is adopted, the cast iron must form the inside of the door, and its convex side must project into the roaster. It should be quite flat near its circumference, in order that it may close the roaster with accuracy; and it should be at least three quarters of an inch larger in diameter than the roaster, in order that no part of the circular plate of sheet iron, which should be fastened to it by being turned over its edge, may get between it and the end of the roaster.

## Of the Blowpipes.

There are various ways in which the blowpipes may be fastened to the roaster. The common method, when they are made of sheet iron, is to fasten them with rivets; but as blowpipes of sheet iron are liable to be burned out in a few years, if much used, it is better to procure them of cast iron from an iron founder, in which case they should be cast with flanges, and should be keyed on the inside of the roaster; and their joinings with the bottom of the roaster must be made tight with some good cement that will stand fire, and is proper for that use.

The effect of the blowpipes will be considerably increased if a certain quantity of iron wire, in loose coils, or of iron turnings, be put into them. These being heated by the fire, the air which passes through the tubes, coming into contact with them, will be more heated than it would be if the tubes were empty; but care must be taken that the quantities of these substances used be not so great as to choke up the tube and obstruct too much the passage of the air.

The stoppers of the blowpipes must be made to close them well, otherwise air will find its way through the blowpipes into the roaster at times when it ought not to be admitted. One of these stoppers, represented on a large scale, is seen drawn a little way out of its

blowpipe, in the Fig. 17, page 183; and in that figure part
of the iron strap is seen which supports the front ends of
the two blowpipes, and confines them in their places.
This strap will not appear when the roaster is set, for
it will then be entirely covered and concealed by the
brick-work.

Where blowpipes are made of sheet iron, they should
be so constructed and so fastened to the roaster that
they may at any time be removed and replaced without taking the roaster out of the brick-work.    This
is necessary, in order that they may be taken away to
be repaired or replaced with new ones, when by long
use they become burned out and unfit for service.    If
they be made with flanges, and keyed on the inside, and
if they be supported in front on an iron strap of the
form represented in Fig. 14, page 153, they may at any
time be removed with little trouble, by unkeying them
and removing a few bricks.    When the bricks in front,
which it will be necessary to take away, are removed,
this will open a passage into the fire-place sufficiently
large to come at the wall at the farther end of the
fire-place, which must come away in order to disen-
gage the farther ends of the blowpipes, which are fixed
in it.    This wall must be carefully built up again, after
the new blowpipes have been introduced and fastened
to the roaster.

## Of the Steam-tube.

This is an essential part of the machinery of a roaster,
and must never be omitted.    It should be situated some-
where in the upper part of the roaster, but it is not
necessary that it should be placed exactly at the top of
it.    It might perhaps be thought that a hole in the

upper part of the door would serve the purpose of a steam-tube ; but this contrivance would not be found to answer. A steam-tube, properly constructed, will have what is called *a draught* through it, which on some occasions will be found to be very useful ; but a hole in the door unconnected with a tube could have no draught. It is absolutely necessary that there should be a damper in the steam-tube. The simplest damper is a circular plate of iron, a very little less in diameter than the tube, which, being placed in it, is movable about an axis, which is perpendicular to the axis of the tube. This circular plate being turned about, and placed in different positions in the tube by means of its axis, which, being prolonged, comes forward through the brick-work, the passage of the steam through the tube is more or less obstructed by it. This prolonged axis, which may be called the projecting handle of this damper, is represented in the Figs. 14, 15, and 17. This appears to me to be one of the simplest kind of dampers I am acquainted with ; and it has this in particular to recommend it, that it may be regulated without opening any passage into the steam-tube, or into the roaster, by which the air could force its way.

### Of the Dripping-pan.

As the principal dripping-pan of a roaster is destined for holding water, and as it is of much importance that it should not leak, it should be hammered out of one piece of sheet iron, in the same manner as a frying-pan is formed ; or, if the metal be turned up at the corners, it should be lapped over, but not cut, and all riveting-nails should be avoided, except such as can be placed very near the edge of the pan, and above the common level of the water that is put into it. To avoid the necessity of placing any riveting-nail at the bottom of the pan or near it, in fastening the sliders on which the pan runs, these sliders should be made to pass upwards by the ends of the pan, in order to their being fastened to it near its brim.

The dripping-pan should not be made quite so long as the roaster, for room must be left between the farther end of it and the farther end of the roaster for the hot air from the blowpipes to pass up into the upper part of the roaster. In order to stop the dripping-pan in its proper place when it is pushed into the roaster, the farther end of the shelf on which it slides may be turned upwards, and the brim of the dripping-pan made to strike against this projecting part of the shelf. The opening between this projecting part of the shelf and the farther end of the roaster should be about 1 inch or $1\frac{1}{4}$ inches wide, and it may be just as long as the dripping-pan is wide at the brim. This part of the shelf which projects upwards should be $\frac{1}{2}$ an inch higher than the brim of the dripping-pan, in order to prevent the current of hot air from the blowpipes from striking against the end of the dripping-pan, and heating it too much. The shelf may be stopped in its proper place by means of two horizontal projecting slips of iron about 1 inch or $1\frac{1}{4}$ inches long each at its farther end, which, striking against the end of the roaster, will prevent the shelf from being pushed too far into it. The dripping-pan should have two falling handles, one at each end of it, which handles should have stops to hold them fast when they are raised into a horizontal position. As these handles will necessarily project a little beyond the ends of the pan, even when they are not raised up, the handle at the farther end of the pan will prevent the brim of the pan from actually touching the projecting end of the shelf; which circumstance will be advantageous, as it will serve to defend the end of the pan, and prevent its being so much heated as otherwise it would be by the hot air from below.

I find, on inquiry from several persons who have lately made the experiment, that it is by far the best method to use two dripping-pans, one within the other, with water between them.  As the upper pan is very thin, being made of tin * (tinned sheet iron), it is kept as cool as is necessary by the water; and, the surface of the water being covered and protected, it does not evaporate so fast as when it is left exposed to the hot air in the roaster.

*Of the Precautions that may be used to prevent the Dripping-pan from being too much heated.*

This is a very important matter, and too much attention cannot be paid to it by those who construct roasters. From what has been said, it is evident that, if in roasting meat the water in the dripping-pan ever happens to be all evaporated, the drippings from the meat which fall on it cannot fail to fill the roaster with noxious fumes. It is certainly not surprising that those who, in roasting in a roaster, neglected to put water into the dripping-pan should not much like the flavour of their roasted meat.

There is a method of defending the dripping-pan from heat, which many have put in practice with success;

---

* Some persons have used a shallow earthen dish, instead of this second dripping-pan; but earthen-ware does not answer so well for this use as tin, as it is more liable to be heated too much by the radiant heat from above.

but, although it effectually answers the purpose, yet it is attended with a serious inconvenience, which, as it is not very obvious, ought to be mentioned. When the bottoms of roasters were made flat, their dripping-pans were much more liable to be too much heated than they are when, the body of the roaster being made cylindrical, the dripping-pan is placed on a shelf in the manner I have here recommended. And several persons, finding the water in the dripping-pans of their roasters to boil away very fast, covered the (flat) bottoms of their roasters with sand, or with a paving of thin tiles or bricks. This produced the desired effect; but this contrivance occasions the bottom of a roaster to be very soon burned out and destroyed. The heat from the fire communicated to the under side of the bottom of the roaster, not being able to make its way upwards into the body of the roaster through the stratum of sand or bricks (which substances are non-conductors of heat), it is accumulated in the bottom of the roaster, and becomes there so intense as to destroy the iron in a short time.

The best method that can be adopted for preventing the dripping-pan from being too much heated is to defend the bottom of the roaster from the direct action of the fire by interposing a screen of some kind or other between it and the burning fuel. This screen may be a plate of cast iron, about one third of an inch thick, with a number of small holes through it, supported upon iron bars at the distance of about an inch below the bottom of the roaster; or it may be formed of a row of thin flat tiles laid upon the blowpipes, and supported by them.

Roasters which are made of a cylindrical form will hardly stand in need of any thing to screen them from the fire, especially if their fire-places are situated at a proper distance below them, and if the size of the fire is kept within due bounds. But, after all, if the person to whom the management of a roaster is committed is determined to destroy it, no precautions can prevent it; and hence it appears how very necessary it is to secure the good-will of the cooks. They ought certainly to wish well to the success of these inventions; for the introduction of them cannot fail to diminish their labour, and increase their comforts very much.

## *Of large Roasters, that will serve to roast and bake at the same Time.*

It has been found by experience that any roaster may be made to roast and bake at the same time, in great perfection, when the proper precautions are taken ; but this can best be done when the roaster is of a large size, from 20 inches to 24 inches in diameter, for instance; for in this case there will be room above the meat for a shelf on which the things to be baked can be placed. And even when there is no roasting going on below it, any thing to be baked should be placed on this shelf, in order to its being nearer to the top of the roaster, where the process of baking goes on better than anywhere else. In baking bread, pies, cakes, etc., it seems to be necessary that the heat should descend in rays from the top of the oven; and as the intensity of the effects produced by the calorific rays which proceed from a heated body is much greater near the hot body than at a greater distance from it (being most probably as the squares of the distances inversely), it is evident why the process of baking should go on best in a low oven, or when the thing to be baked is placed near the top of the oven, or of the roaster, when it is baked in a roaster.

The shelf in the upper part of a roaster for baking may be made of a single piece of sheet iron, but it will be much better to make it double; that is to say, of two pieces of sheet iron, placed at a small distance from each other, and turned inwards, and fastened together at their edges, in the manner which will presently be more particularly described. This shelf, whether it be made single or double, should be placed upon ledges, riveted to the sides of the roasters; and, to prevent the hot air from the blowpipes from passing up between the farther end of this shelf and the farther end of the roaster, the shelf should be pushed quite back against the end of the roaster. It should be made shorter than the roaster by about two inches, in order that there may be sufficient room, between the hither end of the shelf and the inside of the door of the roaster, for the vapour that ought to be driven out of the roaster to pass upwards to the opening of the steam-tube. This shelf should not be fastened in its place, for it may sometimes, when very large pieces of meat are roasted, be found necessary to remove it.

As it seems probable that radiant heat from the top and sides of the roaster acts an important part, even in the process of roasting, if a roaster of very large dimensions were to be constructed, I think it would be advisable not to make its transverse section circular, but elliptical, the longest axis of the ellipse being in a horizontal position. This form would bring the top of the roaster to be nearer to the meat than it would be if its form were cylindrical, its capacity remaining the same. How far a horizontal shelf of sheet iron, placed immediately over the meat, and very near it, would answer as a remedy for the defect of a roaster, the top of which, on account of its great size, should be found to be too far from the surface of the meat, I cannot pretend to determine, as I never have made the experiment; but I think it well deserving of a trial. If the farther end of this shelf were made to touch the farther end of the roaster, so as to prevent the current of air from the blowpipes from getting up between them, it is very certain that this hot air would be forced to impinge against the shelf, and run along the under side of it, to the hither end of the roaster. The only question remaining, and which can only be determined by experiment, is whether this hot air would heat the shelf sufficiently, or to that temperature which is necessary in order that the iron may throw off those calorific rays which are wanted.

If this shelf were covered above with a pavement of tiles, or if it were constructed of two sheets of iron placed parallel to each other, at the distance of about one inch, turned in or made dishing at their edges, and seamed together at their ends and sides in such a manner as to confine the air shut up between them, either of these contrivances, by obstructing the heat in its passage through the shelf, would promote its accumulation at its under surface, which would not only increase the intensity of the radiant heat where it is wanted, but, by diminishing the quantity of heat which passes through the shelf, would be very useful when any thing is placed on it in order to be baked.

Whenever a shelf is made in a roaster, whether it be situated above the dripping-pan or below it, I think it would always be found advantageous to construct it in the manner here described, viz., of two sheets of iron, with confined air between them; or perhaps it may be still better to fill this cavity with finely pulverized charcoal. The additional expense of constructing the shelves of roasters in this manner would be but trifling; and the passage of the heat through them, which it is always desirable to prevent as much as possible, will, by this simple contrivance, be greatly obstructed. If the lower shelf be so constructed, it will no doubt be found very useful in preventing the too quick evaporation of the water in the dripping-pan.

*Of various Alterations that have been made in the Forms of Roasters, and of the Advantages and Disadvantages of each of them.*

The blowpipes of all the roasters that were constructed, till very lately, were made to pass round to

Fig. 19.

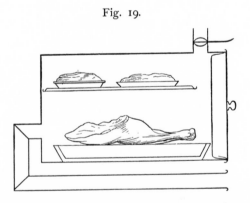

the farther end of the roaster; and, after forming two right angles each, they entered the roaster, in a horizontal direction, just above the level of the brim

of the dripping-pan, in the manner represented in the
Fig. 19.

The Fig. 20 shows the manner in which the blowpipes have been constructed of late.

Fig. 20.

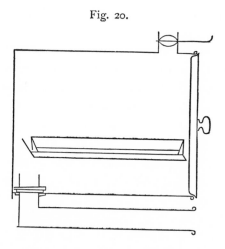

The advantages of the former construction were a
great length of tube, and consequently a greater effect
on that account; and a good direction to the current
of hot air. The disadvantages were the difficulty of
removing the tubes to repair them, without unsetting
the roaster; and the difficulty of procuring blowpipes
of this form of cast iron; and, lastly, the great depth
of space that was required for setting the roaster.

The advantages of the blowpipe, represented in
Fig. 20, have already been noticed. The disadvantage
from want of length is compensated by a small increase
of diameter. When this blowpipe is fastened to the
roaster, its flange is covered with a cement; and
the vertical end of the pipe being introduced into
the roaster through the circular hole in the bottom

of it, which is made to receive it, a flat iron ring, covered with cement on its under side, is then slipped over the end of the tube within the roaster, and a key of iron, in the form of a wedge, being passed through both sides of the tube in holes prepared to receive it, by driving this wedge-like key with a hammer, the ring is forced downwards, and at the same time the flange of the blowpipe is forced upwards against the bottom of the roaster, by which means the blowpipe is firmly fixed in its place, and the cement makes the joinings air-tight. By removing this key, the pipe may at any time be removed without deranging the roaster.

The Fig. 19 represents the section of a flat-bottomed roaster. In this there is a shelf on which two pies are seen baking, and a piece of meat is represented lying on the gridiron.

In the Figs. 14 and 15, pages 153, 157, the front or hither end of the roaster is represented as being turned over a stout iron wire. The first roasters that were constructed were all made in a different manner. The hither end of the roaster was riveted to a broad flat frame, constructed of stout plate iron; and to this frame, or flat front, which projected before the brick-work, the hinges and turn-buckles of the door were fastened. An idea of this manner of constructing the front of a roaster may be formed from the Fig. 21, page 206, although this figure does not represent the front of a roaster, but that of an oven, which will be described presently.

There is no objection to this method of constructing roasters but the expense of it.

*Of some Attempts to simplify the Construction of the Roaster.*

Finding that much more heat was always communicated to the under sides of roasters, especially as they were first constructed (with flat bottoms), than was there wanted, meditating on the means I could employ to defend the bottom of the dripping-pan from this excessive heat, without at the same time exposing the bottom of the roaster to the danger of being soon destroyed, in consequence of the accumulation of it on its passage upwards being prevented, it occurred to me that if the bottom of the roaster were covered with a shallow iron pan turned upside down, with a row of holes from side to side at the farther end of it, and if a certain quantity of fresh air could occasionally be admitted under this inverted pan, this cold air, on coming into contact with the bottom of the roaster, would take off the heat, and, becoming specifically lighter on being heated, would pass upwards through the holes at the farther end of this pan into the roaster, serving at the same time three useful purposes; namely, to defend the dripping-pan; to cool the bottom of the roaster; and to assist in heating the inside of the roaster above, where heat is most wanted. This invention was put in practice, and was found to answer very well all the purposes for which it was contrived. It was likewise found that with proper management the current of heated air from below the inverted pan might be so regulated as to roast meat very well without making any use of the blowpipes; and consequently that roasters might be constructed without blowpipes.

As the substitution of the contrivance above described, in lieu of the blowpipes, would simplify the
construction of the roaster very much, and enable
tradesmen to afford the article at a much lower price,
I took a great deal of pains to find out whether a
roaster on this simple construction could be made to
perform as well as those which are made with blowpipes. I caused one of them to be put up in my own
house, and tried it frequently; and I engaged several
of my friends to try them; and they were found to
answer so well that I ventured at length to recommend it to manufacturers to make them for sale. As
they were called roasters, and as they cost little more
than half what those with blowpipes were sold for, many
persons preferred them on account of their cheapness;
and more than two hundred of them have already been
put up in different parts of the country, and I am informed that they have answered to the entire satisfaction of those who have tried them.

Although they are undoubtedly inferior in some respects to roasters which are furnished with blowpipes,
meat may, with a little care and attention, be roasted
in them in very high perfection; and, as nothing can
possibly answer better than they do for all kinds of
baking, they will, I am persuaded, find their way in
due time into common use.

Roasters on this simple construction (without blowpipes), which I shall call *Roasting Ovens*, were at first
made with flat bottoms, but of late they have been made
cylindrical; and, as I think the cylindrical form much
the best in many respects, I shall give a description of
one of them.

Fig. 21 represents a front view of a cylindrical

roasting oven with its door shut. The front end of the
large cylinder, which constitutes the body of this oven,
instead of being turned over a stout wire, is turned outwards, and riveted to a flat piece of thick sheet iron,
which in this figure is distinguished by vertical lines,
and which I shall call *the front* of the oven.

Fig. 21.

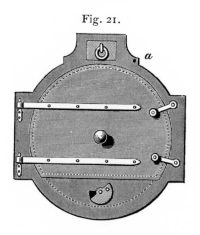

The door of the oven is distinguished by horizontal
lines. The general form of the front of the oven is
circular; but it has two projections on opposite sides
of it, to one of which the hinges of the door, and to the
other the turn-buckles for fastening it when it is closed,
are fastened. It has another projection above, which
serves as a frame to the doorway, through which a
brush is occasionally introduced for the purpose of
cleaning the flues. On one side of this projection
there is a small hole, which is distinguished by the
letter *a*, through which the handle or projecting axis
of the circular register of the vent-tube (which is not
seen) passes.

In the body of the oven, at the distance of half its

semi-diameter below its centre or axis, there is a horizontal shelf, which is fixed in its place, not by resting on ledges, or by being riveted to the sides of the oven, but by its hither end being turned down, and firmly riveted to the vertical plate of iron, which I have called the front of the oven. This shelf, which should be made double to prevent the heat from passing through it from below, must not reach quite to the farther end of the oven : there must be an opening left, about one inch in width, between the end of it and the farther end of the oven, through which opening the air heated below the shelf will make its way upwards into the upper part of the oven.

From what has been said, it will be evident that the hollow space below the shelf we have just been describing, which I shall call the *air-chamber*, is intended to serve in lieu of the blowpipes of a roaster; and this office it will perform tolerably well, provided means are used for admitting cold air into it, from without, occasionally. This is done by means of a register, which is situated at the lower part of the vertical front of the roaster, a little below the bottom of the door. This register is distinctly represented in the Fig. 21.

Fig. 22, which represents a vertical section of the oven through its axis, shows the (double) door of the roaster shut, and the two dripping-pans, one within the other, standing on the shelf we have just been describing, and a piece of meat above them, which is supposed to be laying on a gridiron placed in the second dripping-pan. The register of the air-chamber below the shelf, which supplies the place of the blowpipes, is represented as being open; and a part of the steam-tube is shown, through which the steam and vapour are driven out of the oven, by the blast of hot air from the air-chamber.

The cylinder which constitutes the body of the oven is two feet long, and is supposed to be of cast iron. It is cast with a flange, which projects outwards about one inch at the opening of the cylinder, by means of which flange it is attached, by rivets, to the front of the oven, which, as I have already observed, must be made of strong sheet iron, which may be near one eighth of an inch in thickness.

Fig. 22.

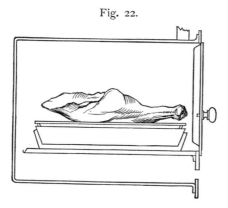

As the shelf is not attached to the sides of the oven, but to its front, the body of the oven need not be perforated, except in one place, namely, where the steam goes off; and as the bottom or farther end of the cylinder, and the flange at its hither end, and the cylinder itself, are all cast at the same time, and as the form of the oven is such as will deliver well from the mould, it appears to me that the article might be afforded at a low price, especially in this country, where the art of casting in iron is carried to so high a pitch of perfection.

The shelf might easily be made of cast iron, as might also the dripping-pans and the double door of the oven; and I should not be surprised if English workmen should succeed in making even the front of the oven and the register of the air-chamber, and every other part of the machinery, of that cheap and most useful metal.

If the shelf be made of cast iron, to save the trouble of riveting in making it double, it may be covered by an inverted shallow pan of cast iron; and in the bottom of this pan, which will be uppermost when it is inverted, there may be cast two shallow grooves, both in the direction of the length of the pan, and consequently parallel to each other, in which grooves (which may be situated about an inch from the sides of the inverted pan) two parallel projections at a proper distance from each other, cast at the bottom of the lower dripping-pan, may pass. These projections, passing freely in the grooves which receive them, will serve to keep the dripping-pan steady in its proper direction when it is pushed into or drawn out of the oven.

To increase the effect of the air-chamber when this oven is used for roasting meat, a certain quantity of iron wire in loose coils, or of iron turnings, may be put into the air-chamber.

The door of the oven, which is very distinctly represented in the Fig. 21, should be about 19 inches in diameter, if the oven is 18 inches in diameter within, or in the clear. In this figure the internal edge or corner of the hither end of the body of the oven is indicated by a dotted circle, and the position of the shelf is pointed out by a horizontal dotted line.

In fastening the vertical plate, which forms the front of the oven, to the projecting flange at the hither end of the cylindrical body of the oven, care must be taken to beat down the heads of the riveting nails in front, otherwise they will prevent the door of the oven from closing it with that nicety which is requisite.

In setting this roasting-oven, the whole of the thickness of the vertical front of it should be made to project forward before the brick-work. The fire-place doors, ash-pit, register-door, damper in the chimney, etc., should be similar in all respects to those used for roasters; and the flues should likewise be constructed in the same manner.

I have been the more particular in my description of this roasting-oven, because I think it bids fair to become a most useful implement of cookery. As an oven, it certainly has one advantage over all ovens constructed on the common principles, which must give it a decided superiority. By means of the air-chamber and the steam-tube it may be kept clear of all ill-scented and noxious fumes without the admission of cold air.

### *Of the Difference between a Roasting-oven and a Roaster.*

From the account of the roasting-oven that has just been given it might be imagined that it possesses all the properties of the roaster, and in the same degree; but this is not the case. The essential difference between them is this: the blowpipes of the roaster being surrounded by the flame on all sides, they are heated above as well as below, and the air in passing through them is much more exposed to the heat than it is in passing through the air-chamber of the roasting-oven. The particles of air which happen to come into contact with the bottom of the oven will of course be heated;

but if, in consequence of their acquired lightness on being heated, they rise upwards to the top of the air-chamber, they will there come in contact with the bottom of the shelf, which, instead of communicating more heat to them, will deprive them of a part of that which they bring with them from below.   But circumstances are very different in the blowpipes of a roaster: in them the particles of air acquire continually additional heat from every part of the surface with which they come into contact in their passage through the tube.

From this view of the subject, we see how very essential it is that the shelf of a roasting-oven should be so composed or constructed that heat may not readily find its way through it; and we see likewise how necessary it is to manage the registers of blowpipes and of air-chambers with proper care.

---

## CHAPTER  VI.

*Of the Usefulness of small iron Ovens, and of the best Methods of constructing them and managing them. — Reasons why they have not succeeded in many Cases where they have been tried. — Ovens may be used for other Processes of Cookery besides Baking. — Curious Results of some Attempts to boil Meat in an Oven. — Explanation of these Appearances. — Conjectures respecting the Origin of some national Customs.*

IN the first part of this tenth Essay I recommended small iron ovens for cottagers, and nests of small ovens for the kitchens of large families; and I have had occasion to know since that several persons have adopted them. I have likewise been made acquainted with the results of many of the trials that have been made of them, and with the complaints that have been brought against them. As I am more than ever of opinion that iron ovens will always be found useful when they are properly constructed and properly managed, I shall in this place add a few observations to what I have already published concerning them.

And, in the first place, I must observe that a small iron oven stands in need of a good door; that is to say, of a door well contrived for confining heat; and the smaller the oven is, so much the more necessary is it that the door should be good.

The door must not only fit against the mouth of the oven with accuracy, but it must be composed of materials through which heat does not easily make its way.

An oven door constructed of a single sheet of plate iron will not answer, however accurately it may be made to fit the oven; for the heat will find its way through it, and it will be carried off by the cold air of the atmosphere which comes into contact with the outside of it. The bottom of the oven may be made hot by the fire under it; but the top and sides of it cannot be properly heated while there is a continual and great loss of heat through its door. But an oven, to perform well, must be very equally heated in every part of it.

If the flame and smoke of the fire be made to surround an oven on every side, and if the fire be properly managed, there can be no difficulty in heating an iron oven equally, and of keeping it at an equal temperature, provided the loss of heat by and through the door be prevented.

If the door be constructed of sheet iron, it must either be made double, or it must be covered on the outside with a panel of wood.   By a *double door* I do not here mean *two doors*, but one door constructed of two sheets or plates of iron placed parallel to and at a certain distance from each other; and so constructed that the air which is between the two plates may be shut up and confined. The two plates or sheets of iron, of which the double door of an oven is made, must not touch each other, except at their edges (where they must join in order to their being fastened together); for, were they to lie one flat upon the other, the heat would pass too rapidly through them, notwithstanding there being two of them; but it is not necessary that they should be farther asunder than an inch or an inch and a half.   One of the plates may be quite flat, and the other a little convex.   The end of the oven must be made quite flat or level, so as to be perfectly closed by a flat surface placed against it.   The door is that flat surface; and the greatest care must be taken that it apply with accuracy, or touch the end of the oven in every part when it is pressed against it; for if any opening be left, especially if it be near the top of the oven, the hot air in the oven will not fail to make its escape out of it.

It never should be attempted to make the door of an oven or of a closed fire-place fit, by causing it to *shut into a rabbet*.   That is a very bad method; for, besides the difficulty of executing the work with any kind of accuracy, the expansion of the metal with heat is very apt to derange the machinery, when the door is so constructed.

From what has been said of the necessity of causing the door of an oven to fit with accuracy, it is evident that care must be used to place its hinges properly; and I have found, by experience, that such a door is closed more accurately by two turn-buckles, placed at a proper distance from each other, than by a single latch. I beg pardon for repeating what has already been said elsewhere.

## *Of the Management of the Fire in heating an iron Oven.*

If a certain degree of attention is always necessary in the management of fire, there is certainly nothing on which we can bestow our care that repays us so amply; and, with regard to the trouble of managing a fire in a closed fire-place, it is really too inconsiderable to deserve being mentioned.

Whenever a fire is made under an iron oven, in a closed fire-place, constructed on good principles, there is always *a very strong draught* or pressure of air into the fire-place; and this circumstance, which is unavoidable, renders it necessary to keep the fire-place door constantly closed, and to leave but a small opening for the passage of the air through the ash-pit register. The fire-place, too, should be made very small, and particularly the bottom of it, or the grate on which the fuel burns.

If any of these precautions are neglected, the consequences will be, — the rapid consumption of the fuel, the sudden heating and burning of the bottom of the oven, and the sudden cooling of the oven as soon as the fire-place ceases to be filled with burning fuel.

It is a fact which ought never to be forgotten, "that of the air that forces its way into a closed fire-place, that part only which comes into actual contact with the burning fuel, and is decomposed by it in the process of combustion, contributes any thing to the heat generated; and that all the rest of the air that finds its way into and *through* a fire-place is a thief that steals heat, and flies away with it up the chimney."

The draught occasioned by a fire in a closed fire-place being into the chimney and not into the fire, cold air is as much disposed to rush in over the fire as through it; and it violently forces its way into the hot fire-place by every aperture, even after all the fuel is consumed, carrying the heat away with it up the chimney and into the atmosphere. It even makes its way between the bars of the grate whenever they are not quite covered with burning fuel; hence it appears how necessary it is to make the grate of a closed fire-place small, and to give to that part of the fire-place which is destined for holding the fuel the form of an inverted truncated cone or pyramid, or else to make it very deep in proportion to its length and width.

But the prevention of the air from finding its way through the fire-place without coming into contact with the burning fuel is not the only advantage that is derived from constructing closed fire-places in the manner here recommended: it serves also to increase the intensity of the heat in that part of the fire-place which contains the fuel, which tends very powerfully to render the combustion of the fuel complete, and consequently to augment the quantity of heat generated in that process.

To prevent the bottom of the oven (or boiler) from being too much affected by this intense heat, nothing more is necessary than to make the fire-place sufficiently small, and to place it at a sufficient distance below the bottom of the oven. It will be indispensably necessary, however, with such a (small) fire-place, situated far below the bottom of an oven, to keep the fire-place door well closed, otherwise so much cold air will rush in over the fire that it will be quite impossible to make the oven hot.

I have found by recent experiments that a fire-place in the form of an oblong square or prism, 6 inches wide, 9 inches long, and 6 inches deep, is sufficient to heat an iron oven 18 inches wide, 24 inches long, and from 12 to 15 inches in height; and that the grate of this fire-place should be placed about 12 inches below the bottom of the oven. More effectually to prevent the fire from operating with too much violence upon any one part of the bottom of the oven, the brick-work may be so sloped outwards and upwards on every side from the top of the burning fuel to the extreme parts of the sides and ends of the bottom of the oven, that the whole of the bottom of the oven may be exposed to the direct rays from the fire.

In some cases I have suffered the flame to pass freely up both sides of the oven to the top of it, and then caused it to descend by the end of the oven to the level of its bottom, or rather below it, and from thence to pass off by a horizontal canal into the chimney; and in other cases I have caused it to pass backwards and forwards in horizontal canals by the sides of the oven, before I permitted it to go off into the chimney. Either of these methods will do very well, provided the smoke be made to descend after it has left the top of the oven, till it reaches below the level of the bottom of it, before it is permitted to pass off into the chimney; and provided the canal by which the smoke passes off be furnished with a damper.

In setting an oven, provision should be made, by leaving holes to be stopped up with stoppers, for occasionally cleaning out all the canals in which the smoke is made to circulate; and, in order that these canals may not too often be choked up with soot, they should never be made less than two inches wide, even where they are very deep or broad; and, where they are not more than four or five inches deep, they should be from three to four inches wide, otherwise they will be very often choked up with soot.

To clean out the flues of an oven, roaster, or large fixed boiler, a strong cylindrical brush may be used, which may have a flexible handle made of three or more iron wires, about $\frac{1}{8}$ or $\frac{1}{10}$ of an inch in diameter, twisted together.

Holes closed with fit stoppers must of course be left in the brick-work for occasionally cleaning out these flues.

If the iron door of an oven be made double, the outside of it may with safety be japanned black or white, which will prevent its rusting, and add much to the cleanliness and neatness of the appearance of the kitchen.

These details may by some be thought unimportant and tiresome, but those who know how much depends on minute details in the introduction of new mechanical improvements will be disposed to excuse the prolixity of these descriptions. I wish I could make my writings palatable to the generality of readers, but that, I fear, is quite impossible. My subjects are too common and too humble to excite their curiosity, and will not bear the high seasoning to which modern palates are accustomed.

A great disadvantage under which I labour is that, of those who *might* profit most from my writings, many *will not read*, and others *cannot*.

But to return to my subject. To save expense, small ovens for poor families may be closed with flat stones or with tiles; and the fire-place door for such an oven, and its ash-pit register, may be made of common bricks placed edgewise, and made to slide against those openings.

There is a circumstance respecting the iron ovens I am describing, which is both curious and important. The fire-place for an oven of the smallest size should be nearly as capacious as one which is destined for heating a much larger oven; and I have found, by repeated experiments, that a nest of four small ovens, set together, and heated by the same fire, will require but very little more fuel to heat them than would be necessary to heat one of them, were it set alone. An attentive consideration of the manner in which the heat is applied — of the smallness of the quantity, in all cases, that is applied to the heating of the contents of the oven, and the much greater quantity that is expended in heating the fire-place and the flues — will enable us to account for this curious fact in a manner that is perfectly philosophical and satisfactory.

A cottage oven 11 inches wide, 10 inches high, and 16 inches long, will require a fire-place 5 inches wide, 5 inches high, and 7 inches long; and for four of these ovens, set together in a nest, the fire-place need not be more than 6 inches wide, 6 inches high, and 8 inches long.

I have in my house at Brompton two iron ovens, each 18 inches wide, 14 inches high, and 24 inches long, set one over the other, and heated by the same fire; and their fire-place is only 6 inches wide, 6 inches high, and 9 inches long.

If the fire-place of an iron oven be properly constructed, and if the fire be properly managed, it is almost incredible how small a quantity of fuel will answer for heating the oven, and for keeping it hot. But if the fire-place door be allowed to stand open, and a torrent of cold air be permitted to rush into the fire-place and through the flues, it will be found quite impossible to heat the oven properly, whatever may be the quantity of fuel consumed under it; and neither the baking of bread nor of pies, nor any other process of cookery, can be performed in it in a suitable manner.

A very moderate share indeed of ingenuity is required in the proper management of a fire in a closed fire-place, and very little attention. And as it requires no bodily exertion, but saves labour and expense and anxiety; and as moreover it is an interesting and amusing occupation, attended by no disgusting circumstance, and productive of none but pleasing, agreeable, and useful consequences, we may, I think, venture to hope that those prejudices which prevent the introduction of these improvements will in time be removed.

It is not obstinacy, it is that *apathy* which follows a total corruption of taste and morals, that is an *incurable* evil; for that, alas! there is no remedy but calamity and extermination.

*Ovens may be used in boiling and stewing, and also in warming Rooms.*

There are so many different ways in which the heat necessary in preparing food may be applied, that it would not be surprising if one should sometimes be embarrassed in the choice of them; and I am not without apprehension that I may embarrass my readers by describing and recommending so many of them. The fact is, they all have their different kinds of merit, and in the choice of them regard must always be had to the existing circumstances.

Desirous of contriving a fire-place on as simple a construction as possible, that should serve at the same time for heating a room and for the performance of all the common processes of cookery for a small family, and which moreover should not be expensive nor require much attendance, I caused four small iron ovens to be set in the opening of a common chimney fire-place. These ovens, which were constructed of sheet iron, and were furnished with doors of the same sheet iron, each covered with a panel of wood to confine the heat, were 16 inches long, 11 inches wide, and 10 inches high each; and they were set in brick-work in such a manner that the fronts of the doors of the ovens being even with the side of the room, the original opening of the chimney fire-place, which was large, was completely filled up. These ovens were all heated by one small fire, the closed fire-place being situated about 12 inches below the level of the bottoms of the two lowermost ovens, and perpendicularly under the division between them, and the passage into the fire-place was closed by a fit stopper.

From this description, it will not be difficult for any person who has perused the preceding chapters of this Essay to form a perfect idea of this arrangement; and it is equally easy to perceive that, had not the open chimney fire-place in which these four ovens were set

been very large, I should have been under the necessity
of enlarging it, or at least of raising its mantel, in order
to have been able to introduce these ovens, and set them
at proper distances from each other.

I shall now proceed to give an account of the experiments that were made with this fire-place.

My first attempt was to warm the room by means of
it.　A small fire being made in its closed fire-place, its
oven doors were all set wide open, and the room, though
by no means small, soon became very warm.　This
warming apparatus was now, to all intents and purposes, a German stove.　By shutting two of the oven
doors, the heat of the room was sensibly diminished;
and by leaving only one of them open it was found that
a moderate degree of warmth might be kept up even
in cold weather.

As no person in this country would be satisfied with
any fire-place, if in its arrangement provision were not
made for boiling a tea-kettle, I caused a very broad
shallow tea-kettle, with a bottom perfectly flat, to be
constructed of common tin, and, filling it with cold
water, placed it in one of the two lower ovens, and shut
the oven door.　Although the fire under the ovens was
but small, it burned very bright, and the water in the
tea-kettle was soon made to boil.

I was not surprised that the water boiled in a short
time, for it was what I expected; but on removing the
tea-kettle I observed an appearance which did surprise
me, and which indicated a degree of heat in the oven
which I had no idea of finding there.　The handle of
the tea-kettle resembled very much in form the handle
of a common tea-kettle, but, like the rest of the kettle,
was constructed of tin, or, to speak more properly, of
tinned sheet iron.

On removing the kettle from the oven I found that the tin on its handle had been melted, and had fallen down in drops, which rested on the body of the kettle below, where they had congealed, having been cooled by the water in the kettle.

This discovery convinced me that I should not fail of obtaining in these ovens any degree of heat that could possibly be wanted in any culinary process whatever: it showed me likewise that degrees of temperature much higher than that of boiling water may exist in a closed oven in which water is boiling; and it seemed to indicate that all the different culinary processes of boiling, stewing, roasting, and baking might be carried on at the same time in one and the same oven. Subsequent experiments have since confirmed all these indications, and have put the facts beyond all doubt. These facts are certainly curious, and the knowledge of them may lead to useful improvements; for they may enable us to simplify very much the implements used in cookery.

Having found that I could boil water in my small ovens, my next attempt was to boil meat in them. I put about three pounds of beef, in one compact lump, into an earthen pot, and filling the pot to within about two inches of its brim with cold water, I set it in one of the lower ovens, shutting the door of the oven, and keeping up a small steady fire in the fire-place. In about two hours and three quarters the meat was found to be sufficiently boiled; and all those who partook of it (and they were not fewer than nine or ten persons) agreed in thinking it perfectly good and uncommonly savoury. On my guard against the illusions which frequently are produced by novelty, I should have had doubts respecting the reality of those superior qualities ascribed to this boiled beef, had not an uncommon appearance in the water in which it had been boiled attracted my attention. This water, after the meat had been boiled in it, appeared to be nearly as transparent and as colourless as when it was brought from the pump. It immediately occurred to me that this effect could be owing to nothing else but to the state of perfect quiet in which the water must necessarily have been during the greater part of the time it remained in the oven; and, to determine whether this was really the case or not, I made the following decisive experiment.

Having provided two equal pieces of beef from the same carcass, I put them into two stewpans of nearly the same form and dimensions; one of them, which had a cover, being constructed of earthen-ware, while the other, which had no cover, was made of copper.

Into these stewpans I now put equal quantities of water, — with this difference, however, that while the water put into the copper stewpan was cold, that put into the other was boiling hot. A small fire being now made in the fire-place, these two stewpans, with their contents, were introduced into the two lower ovens. The earthen stewpan was set down upon a ten-inch tile, which had previously been placed in the oven to serve as a support for it, in order to prevent the bottom of the stewpan from coming into immediate contact with the bottom of the oven, and the door of that oven was shut; but the copper stewpan was set down immediately on the bottom of its oven, and the door of that oven was left open during the whole time the experiment lasted.

At the end of three hours the stewpans were taken out of the ovens, and their contents were examined.

The appearances were just what I expected to find them. The meat in each of the stewpans was sufficiently boiled, but there was certainly a very striking difference in the appearance of the liquor remaining in the two utensils; and, if I was not much mistaken, there was a sensible difference in the taste of the two pieces of meat, that boiled in the earthen stew-pan being the most juicy and most savoury. The water remaining in this vessel — and little of it had evaporated — was still very transparent and colourless, and nearly tasteless, while the liquor in the copper stewpan was found to be a rich meat broth.

The result of this experiment recalled very forcibly to my recollection a dispute I had had several years before, in Germany, with the cook of a friend of mine, who at my recommendation had altered his kitchen fire-place; in which dispute I now saw I was in the wrong, and, seeing it, felt a desire more easy to be conceived than to be described to make an apology to an innocent person whom I had unjustly suspected of wilful misrepresentation. This woman (for it was a female cook), on being repeatedly reprimanded for sending to table a kind of soup of inferior quality, which, before the kitchen was altered, she had always been famous for making in the highest perfection, persisted in declaring that she could not make the same good rich soup in the new-fashioned boilers (fitted up in closed fire-places, and heated by small fires) as she used to make in the old boilers, set down upon the hearth before a great roaring wood fire.

The woman was perfectly in the right. To make a rich meat soup, the juices must be washed out of the meat, and intimately mixed with the water; and this washing out in boiling must be greatly facilitated and expedited by the continual and rapid motion into which the contents of a boiler are necessarily thrown when heat is applied to one side of it only, especially when that heat is sufficiently intense to keep the liquid continually boiling with vehemence. I ought, no doubt, to have foreseen this; but how difficult is it to foresee any thing! It is much easier to explain than to predict.

If it be admitted that fluids in receiving and giving off heat are necessarily thrown into internal motions in consequence of the changes of specific gravity in the particles of the fluid, occasioned by the alteration of their temperatures, we shall be able to account, in a manner perfectly satisfactory, not only for the appearances observed in the experiments above mentioned, and for the superior richness of the soup made by the Bavarian cook in her boiler, but also for several other curious facts.

When the copper stewpan, containing cold water and a piece of meat, was put into an iron oven, heated by a fire situated below it, as the bottom of the oven on which the stewpan was placed was very hot, the heat, passing rapidly through the flat bottom of this metallic utensil, communicated heat to the lower stratum of the water, which, becoming specifically lighter on being thus heated, was crowded out of its place, and forced upwards by the superincumbent colder and consequently heavier liquid. This necessarily occasioned a motion in every part of the fluid, and this motion must have been rapid in proportion as the communication of heat was rapid; and it is evident that it could never cease, unless all the water in the stewpan could have acquired and preserved an equal and a permanent temperature, which, under the existing circumstances, was impossible; for, as the door of the oven was left open, the upper surface of the water was continually cooled by giving off heat to the cold atmosphere, which, rushing into the oven, came into contact with it; and, as soon as the water was made boiling hot, an internal motion of another kind was produced in it, in consequence of the formation and escape of the steam, which last motion was likewise rapid and violent in proportion to the rapidity of the communication of heat. Hence we see that the water in the copper stewpan must have been in a state of continual agitation from the time it went into the oven till it came out of it; and the state in which this liquid was found at the end of the experiment was precisely that which might have been expected, on a supposition that these motions would take place. Let us now see what, agreeably to our assumed principles, ought to have taken place in the other stewpan.

In this case, its contents having been nearly boiling hot when the stewpan was put into the oven, and the door of the oven having been kept closed, and the stewpan covered with its earthen cover, and the stewpan being moreover earthen-ware, which substance is a very bad conductor of heat, and being placed not immediately on the bottom of the oven, but on a thick tile, every circumstance was highly favourable not only for keeping up the equal heat of the water, but also for preventing it from receiving additional heat so rapidly as to agitate it by boiling. There is therefore every reason to think that the water remained at rest, or nearly so, during the whole time it was in the oven; and the transparency of this fluid at the end of the

experiment indicated that little or none of the juices
of the meat had been mixed with it.

When the Bavarian cook made soup in her own
way, the materials (the meat and water) were put into
a tall cylindrical boiler, and this boiler was set down
upon the hearth against a wood fire, in such a manner
that the heat was applied to *one side only* of the boiler,
while the other sides of it were exposed to the cold air
of the atmosphere; consequently the communication
of the heat to the water produced in it a rapid circu-
latory motion, and, when the water boiled, this motion
became still more violent. And this process being
carried on for a considerable length of time, the juices
of the meat were so completely washed out of it that
what remained of it were merely tasteless fibres; but
when the ingredients for this meat-soup, taken in the
same proportions, were cooked during the same length
of time in a boiler set in a closed fire-place and heated
by a small equal fire, — this moderate heat being applied
to the boiler on every side at the same time, while the
loss of heat at the surface of the liquid was effectually
prevented by the double cover of the boiler, — the internal motions in the water, occasioned by its receiving
heat, were not only very gentle, but they were so divided into a vast number of separate ascending and
descending small currents, that the mechanical effects
of their impulse on the meat could hardly be sensible;
and as the fire was so regulated that the boiling was
never allowed to be at all vehement (the liquid being
merely kept gently simmering) after the contents of
the boiler were once brought to the temperature of
boiling, the currents occasioned by the heating ceased
of course, and the liquid remained nearly in a state of

rest during the remainder of the time that the process of cooking was continued. The soup was found to be of a very inferior quality, but on the other hand the meat was uncommonly juicy and savoury.

These minute investigations may perhaps be tiresome to some readers; but those who feel the importance of the subject, and perceive the infinite advantages to the human species that might be derived from a more intimate knowledge of the science of preparing food, will be disposed to engage with cheerfulness in these truly interesting and entertaining researches; and such readers, and such only, will perceive that it has not been without design that, in chapters devoted to the explanation of subjects the most humble, I have frequently introduced abstruse philosophical researches and the results of profound meditation.

I am not unacquainted with the manners of the age. I have lived much in the world, and have studied mankind attentively, and am fully aware of all the difficulties I have to encounter in the pursuit of the great object to which I have devoted myself. I am even sensible, fully sensible, of the dangers to which I expose myself. In this selfish and suspicious age, it is hardly possible that justice should be done to the purity of my motives; and in the present state of society, when so few who have leisure can bring themselves to take the trouble to read any thing except it be for mere amusement, I can hardly expect to engage attention. I may write, but what will writing avail if nobody will read. My bookseller, indeed, will not be ruined as long as it shall continue to be fashionable to have fine libraries. But my object will not be attained unless my writings are

read, and the importance of the subjects of my investigations are felt.

Persons who have been satiated with indulgences
and luxuries of every kind are sometimes tempted by
the novelty of an untried pursuit.  My best endeavours
shall not be wanting to give to the objects I recommend
not only all the alluring charms of novelty, but also the
power of procuring a pleasure as new, perhaps, as it is
pure and lasting.

How might I exult could I but succeed so far as to
make it fashionable for the rich to take the trouble
to choose for themselves those enjoyments which their
money can command, instead of being the dupes of
those tyrants who, in the garb of submissive fawning
slaves, not only plunder them in the most disgraceful
manner, but render them at the same time perfectly
ridiculous, and fit for that destruction which is always
near at hand when good taste has been driven quite off
the stage.

When I see in the capital of a great country, in the
midst of summer, a coachman sitting on a coach-box
dressed in a thick heavy greatcoat with sixteen capes, I
am not suprised to find the coach door surrounded by
group of naked beggars.

We should tremble at such appearances, did not the
shortness of life and the extreme levity of the human
character render us insensible to dangers while at any
distance, however great and impending and inevitable
they may be.

But to return from this digression.

It is frequently useful, and is always amusing, to trace
the differences in the customs and usages of different
countries to their causes.   The French have for ages

been remarkable for their fondness for soups, and for their skill in preparing them. Now as national habits of this kind must necessarily originate at a very early period of society, and must depend on peculiar local circumstances, may not the prevalence of the custom of eating soup in France be ascribed to the open chimney fire-places and wood fires which have ever been common in that country?

It is certain that in the infancy of society, before the arts had made any considerable progress, families cooked their victuals by the same fire which warmed them. Kitchens then were not known; and the utensils used in cooking were extremely simple, an earthen pot perhaps set down before the fire. We have just seen that, with such an apparatus, soups of the very best qualities would naturally be produced; and it is not surprising that a whole nation should acquire a fondness for a species of food not only excellent in its kind, but cheap, nutritious, and wholesome, and easily prepared.

Had coals been the fuel used in France, it is not likely that soups would have been so generally adopted in that country; for a common coal fire is not favourable for making good soups, although with a little management the very best soups may be made, and every other process of cooking be performed, *in the highest perfection* with any kind of fuel.

When the *science* of cookery is once well understood, or an intimate knowledge is acquired of the precise nature of those chemical and mechanical changes which are produced in the various culinary processes, we may then, and not till then, take measures with certainty for improving the *art* of preparing food. Experience, unassisted by science, may lead, and frequently does lead, to useful improvements ; but the progress of such improvement is not only slow, but vacillating, uncertain, and very unsatisfactory.   On that account, no doubt, it is that men of science have in all ages been respected as valuable members of society.

# PART III.

## CHAPTER VII.

*Of the Construction of Boilers, Stewpans, etc. — Choice of the Material for constructing Kitchen Utensils. — Objections to Copper. — Iron much less unwholesome. — Of the Attempts that have been made in different Countries to cover the Surface of iron Boilers with an Enamel. — Of Earthen-ware glazed with Salt. — Stewpans and Saucepans of that Substance recommended. — Kitchen Utensils of Earthen-ware may be covered and protected by an Armor of sheet Copper. — Wedgewood's Ware unglazed would answer very well for Kitchen Utensils. — Directions for constructing Stewpans and Saucepans of Copper in such a Manner as to make them more durable, and more easy to be kept clean. — These Utensils are frequently corroded and destroyed by the Operation of what has been called the Galvanic Influence. — Of the Construction of Covers for Kitchen Boilers, Stewpans, etc.*

THE choice of the material to be used in constructing kitchen boilers, stewpans, etc., is a matter of so much importance that I cannot pass it over in silence; though I am very sensible that all I can offer on the subject will not be sufficient to remove entirely the various difficulties I shall be obliged to point out.

231

The objects principally to be had in view in the choice of materials to be used in the construction of kitchen utensils are wholesomeness, cheapness, and durability. The material most commonly used for constructing kitchen boilers and saucepans is *copper;* but the poisonous qualities of that metal, and the facility with which it is corroded and dissolved by the acids which abound in those substances that are used as food, has long been known and lamented.  And numerous attempts have been made to prevent its deleterious effects, by covering its surface with tin and with other metallic substances, and with various kinds of varnish and enamel; but none of these contrivances have completely answered the purpose for which they were designed.

The method which has been found to be most effectual is to keep the copper utensils well tinned, or to tin them afresh as often as the copper begins to appear, and this is what is now commonly practised; but still it were to be wished that some good substitute might be found for that unwholesome metal.

*Iron* has often been proposed; and though it is more liable to be corroded even than copper, yet as the rust (oxide) of iron is not poisonous, though it changes the colour of some kinds of food that are cooked in it, and in some cases communicates an astringent taste to them, it is not thought to make food unwholesome.

There is, however, one precaution by means of which the disagreeable effects produced by this metal on food that is prepared in utensils constructed of it may be very much diminished, and indeed in most cases almost entirely prevented, especially when the utensil is made of *cast iron.*  If, instead of scouring the inside of iron boilers and stewpans with sand, and keeping them bright, which notable housewives are apt to do, in order that their kitchen furniture may appear neat and clean, they be simply washed and rinsed out with warm water, and wiped with a soft dishcloth or towel, the surface of the metal will soon become covered with a thin crust or coating of a dark brown colour resembling enamel; which covering, if it be suffered to remain and to consolidate, will at last become so hard as to take a very good polish, and will serve very efficaciously to defend the surface of the metal from farther corrosion, and consequently to prevent the food from acquiring that taste and colour which iron is apt to impart to it.

The process by which this covering is gradually formed is similar to that by which some gunsmiths brown the barrels of fowling-pieces, and could no doubt be greatly expedited by the same means which they employ for that purpose. The object had in view is likewise the same in both cases, namely, by causing a hard and impenetrable covering of rust to be formed on the surface of the iron to defend it from a contact with those substances which are capable of dissolving or corroding it; or, in other words, to prevent the farther progress of the rust.

For iron utensils designed merely for *frying* or cooking in fat there is an easy and a very effectual precaution that may be taken for preventing rust. It is to avoid putting hot water into them, and above all to avoid boiling, or even heating, water in them. They may occasionally be washed out with warm water; but as often as this is done great care must be taken to wipe them perfectly dry with a dry cloth before they are put away.

The effects produced by this management may be explained in a satisfactory manner. As fatty or oily substances cannot communicate oxygen to iron (with which that metal must unite in order that rust may be formed), and as they prevent the approach of other substances which could furnish it (air, water, acids, etc.) as long as the surface of the iron is completely covered by them, it is evident that no rust can be formed. But boiling-hot water, and more especially water heated and actually made to boil in such a vessel, could not fail to dislodge the fat from the surface of the metal, and leave it naked and exposed to every thing that is capable of corroding it.

Kitchen utensils made of iron may be tinned on the inside to preserve them from rust; and this is frequently done. But even tin, though it be much less liable to be dissolved by those substances which are used in cookery than iron or copper, yet it is sometimes sensibly corroded by them, and consequently is taken into the stomach with our food.

What its effects may be on the human body, when taken in very small quantities, I cannot pretend to determine. In large doses it is well known to be a fatal poison.

That the tin with which the insides of kitchen boilers and stewpans are covered is actually corroded in many of the processes of cookery is rendered highly probable by the very short time that such a coating lasts, when the utensil is in daily use; but I had, not long since, a still more striking proof of that fact. Learning by accident, from my cook, that a dish of which I am very fond (*stewed pears*, which I frequently eat with bread and milk for my supper), required three hours' boiling, it occurred to me that, as this process was performed in a copper stewpan tinned, and as it lasted so long a time, the tin might perhaps be attacked, and some part of it dissolved by the acid of the pears, or by that of the sugar which was mixed with them. In order that I might be able to enjoy my favourite dish free from all apprehensions of being poisoned, I ordered it to be always prepared in future in a stewpan of porcelain; but, several of these vessels having been destroyed in a short time by the fire in this process, I found myself obliged to abandon this scheme on account of these frequent accidents; and I now had recourse to my roaster.

The pears, being previously cut in quarters, and freed from their skins, seeds, and cores, were put, with a sufficient quantity of water and sugar, into a shallow glass basin fitted with a glass cover, and this basin, being placed upon a brick, was put into the roaster; and, a small fire being made under it, the water in the basin was soon brought to boil, and in less than three hours the pears were found to be sufficiently done:

When they were served up, I observed that their colour was different from what it had always been before; and, inquiring into the cause of it, I was let into a secret which explained the matter completely. The cook informed me that it was absolutely impossible to give *a beautiful red colour* to stewed pears without some metal, and that their colour would not have been so fine as it was when they were cooked in porcelain, had not the precaution been taken *to boil a pewter spoon with them.* The reader can easily imagine how much I was surprised at receiving this unexpected information.

This ingenious contrivance is similar to one sometimes used in this country, — that of boiling *half-pence* with greens to give them a fine colour.

Several years ago a variety of attempts was made in Sweden to improve cooking utensils made of iron, by covering them on the inside with a kind of enamel, to protect them from rust; and since that time a considerable manufacture of cast iron boilers and stewpans, covered within with white enamel, was established by Count Heinitz, on his estate in Silesia; but this scheme has not succeeded entirely, owing to the difficulty of finding an enamel capable of uniting with iron, the expansion of which with heat shall be so nearly equal to the expansion of iron as not to be liable to crack and fly off upon being suddenly exposed to heat and to cold; and even were it possible to compose an enamel that would withstand the effects of the heat and the cold, and the blows to which it would be exposed in the business of the kitchen, there would still remain a very important point to be ascertained, which is whether the matter of which the enamel is composed *is not itself of a poisonous nature,* and whether there is not reason to apprehend that it might communicate its deleterious qualities to the food.

Lead is an essential ingredient in most, if not all, enamels, and as its effects are known to be extremely pernicious to health, under all its various forms, when taken internally, it would be highly necessary to ascertain, by the most rigid experimental investigation, whether the enamel of kitchen utensils contains any lead or other noxious metals or unwholesome substance; and, if this be the case, whether such poisonous substance be liable to be corroded and dissolved, or mixed in any other manner with the food.

It is possible that a poisonous substance may be so fixed, on being mixed and united with other substances, as to render it perfectly insoluble, and consequently perfectly inert and harmless; but still the fact ought to be well ascertained before it is admitted.

A large proportion of the calx of lead enters into the composition of flint glass, yet it is not probable that flint glass ever communicates any thing poisonous to food or drink that is kept in it. But, on the other hand, there is reason to conclude that the glazing of common pottery, which is likewise composed in part of calx of lead, is not equally safe, when earthen vessels covered with it are used as implements of cookery. In some countries the use of such vessels in the processes of boiling and stewing is forbidden by the laws, under severe penalties; and in this country it is not customary to use earthen vessels, so glazed, for preserving pickles, and other substances designed for the use of the table which contain strong acids.

The best glazing for earthen vessels that are to be used in preparing or preserving food is most undoubtedly made with common salt, as this glazing (which appears to be merely the beginning of a vitrification of the earth at the surface of the vessel) is not only very hard and durable, but it is also perfectly insoluble in all the acids and other substances in common use in kitchens, and contains nothing poisonous or unwholesome.

A large proportion of lead enters into the composition of pewter; but it has lately been proved, by many ingenious experiments made to ascertain the fact, that the lead, united to tin and the other metallic substances that are used in composing pewter, is incomparably less liable to be dissolved by acids, and consequently much less unwholesome than when it is pure or unmixed with other metals.  This fact is very important, as it tends to remove all apprehension respecting the unwholesomeness of a very useful compound metal, which, from its cheapness, as well as on account of its durability, renders it peculiarly well adapted for many domestic uses.  It would not, however, be advisable to boil or stew any kind of food, especially such as contain acids, in pewter vessels; nor should acid substances ever be suffered to remain long in them.

The best, or at least the most wholesome, material for stewpans and saucepans is, undoubtedly, earthen-ware glazed with salt.*  Several manufactories of this kind of pottery have lately been established in this country, and one in particular in the King's Road, at Chelsea, which belonged to the late Mrs. Hempel, which is, I believe, now carried on by her sons.  The principal reason why this article has not long since found its way into common use is, no doubt, the brittleness of earthen-ware, and its being so liable to crack on being suddenly exposed to heat or to cold; for, excepting this imperfection, it has every thing to recommend it.  It is perfectly wholesome (when glazed with salt), and is kept clean with little trouble; and things cooked in

---

* Nothing is more pernicious than the glazing of common coarse earthenware.  There is no objection to *unglazed* earthen-ware but its being apt to imbibe moisture, which renders it difficult to be kept clean.  I have lately seen some kitchen utensils of very fine, compact, unglazed earthen-ware, bought at Mr. Wedgewood's manufactory, which I thought very good.  They were made thin, and seemed to stand the fire very well; and, as their surface was very smooth, they were easily kept clean.  I wish that the intelligent gentlemen who direct that noble manufactory would turn their attention to the improvement of an article so nearly connected with the health, comfort, and peace of mind of a great portion of society.  Stewpans of this material, suspended in a cylindric.l armor of sheet iron, would be admirably calculated for the register stoves I shall recommend.  Some of these stoves may be seen in the great kitchen of the Royal Institution.

it are much less liable to be burned to the sides of
the vessel, and spoiled, than when the utensil is formed
of a metallic substance.

There is a very great difference in earthen-ware in
respect to its power of withstanding the heat without
injury, on being suddenly exposed to the action of a
fire, some kinds of it being much less liable to crack
and fly, when so exposed, than others; and, in order
to take measures with certainty for diminishing this
imperfection, we have only to consider the causes from
which it proceeds. Now it is quite certain that the
cracking of an earthen vessel, on its being put over
a fire, is owing to *two* circumstances, — the brittleness
of the substance, and the difficulty or slowness with
which heat passes through it; for it is evident that
neither of these circumstances alone, or acting singly,
would be capable of producing the effect.

As heat expands all solid bodies, if one side of a vessel, composed of a brittle substance, be suddenly heated
and expanded, it must crack, or rather it must cause
the other surface to crack, unless the heat can make
its way through the solid substance of the vessel, and
heat and expand that other surface so expeditiously
as to prevent that accident. Now, as heat passes
through a vessel which is thin sooner than through
one (composed of the same material) which is thicker,
it is evident that the thinner an earthen vessel for
cooking is made, the less liable will it be to receive
injury on being exposed to sudden heat or cold.

I mention sudden cold as being dangerous, and it is
easy to see why it must be equally so with sudden heat.
If a brittle vessel be (by slow degrees) made very hot,
if the heat be equally distributed throughout the whole

of its substance, this heat, however intense it may be, will have no tendency whatever to cause the vessel to crack; for, the expansion being equal at the two opposite surfaces, the tension at those surfaces will be equal also. But, if cold water be suddenly poured into a vessel so heated, its internal surface will be suddenly cooled and as suddenly contracted; and as the external surface cannot contract, being forcibly kept in a state of expansion by the heat, the inside surface must necessarily crack, in consequence of its contraction, and this fracture will make its way immediately through the whole solid substance of the vessel from the inside to the outside surface.

Sudden heat applied to one side or surface of a brittle vessel causes the *opposite* side of it to crack; but sudden cold *causes the side to crack to which the cold is applied.*

By forming distinct ideas of what happens in these two cases, every thing relative to the subject under consideration will be rendered perfectly clear and intelligible.

The *form* of a vessel has a considerable effect in rendering it more or less liable to be cracked and destroyed by sudden heat or cold. All flat surfaces, sharp corners, and inequalities of thickness, should, as much as possible, be avoided. The globular form is the best of all, and next to it are those forms which approach nearest to it; and the thinner the utensil is made, consistent with the requisite strength to resist occasional blows, the better it will be in all respects.

The best composition for earthen-ware for culinary purposes is, I am told, pounded Hessian crucibles, or any kind of broken earthen-ware of that kind, reduced to powder, and mixed with a very small proportion of Stourbridge clay.

The method of glazing this ware with salt is by throwing decrepitated common salt into the top of the kiln, with an iron ladle, through six or eight holes made for that purpose in different parts of the top of the kiln. These holes, which need not be more than four inches in diameter each, may be kept covered with common bricks laid over them.

The salt should not be thrown in till the ware is sufficiently burned and till it has acquired the most intense heat that can be given it; and the holes should be immediately closed as soon as the salt is thrown in. If as much as a large handful of salt be thrown into each hole, that will be sufficient, unless the kiln be very large.

The salt is immediately reduced to vapour by the intense heat, and this vapour expands itself and fills every part of the kiln, and disposes the ware to vitrify at its surface.

I have made several attempts to protect stewpans and saucepans of earthen-ware from danger from sudden heat, and from accidental blows, by covering them on the outside with sheet copper and with sheet iron; and in these attempts I have succeeded tolerably well. Several stewpans covered in this manner may be seen in the kitchen and in the repository of the Royal Institution. As the subject is of infinite importance to the health and comfort of mankind, I wish that some ingenious and enterprising tradesman would turn his attention to it.

As cooking utensils of tinned iron are incomparably less dangerous to health than those which are made of

copper, I have taken considerable pains to get serviceable stewpans and saucepans made of that material.
The great difficulty was to unite durability with cheapness and cleanliness. How far I have succeeded in this
attempt will be seen hereafter.

As it is probable the copper stewpans and saucepans
will continue to be used, at least for a considerable time
to come, notwithstanding the objections which have so
often been made to that poisonous metal, I shall proceed to an investigation of the best forms for those
utensils.

Before I proceed to a consideration of the improvements that may be made in the forms of kitchen utensils, I must bespeak the patience of the reader. It is quite
impossible to make the subject interesting to those who
read merely for amusement, and such would do well
to pass over the remainder of this chapter without
giving it a perusal; but I dare not treat any part of a
subject lightly which I have promised to investigate.
Besides this, I really think the details, in which I am
now about to engage, of no inconsiderable degree of
importance; and many other persons will, no doubt,
be of the same opinion respecting them. The smallest
real improvement of any utensil in general and daily
use must be productive of advantages that are incalcu-
lable. It is probable that more than a million of kitchen
boilers and stewpans are in use every day in the United
Kingdom of Great Britain and Ireland; and the providing and keeping kitchen furniture in repair is a heavy
article of expense in housekeeping. I am certain that
this expense may be considerably lessened; and, in doing
this, that kitchen utensils may be made much more convenient, neat, and elegant than they now are.

As it is indispensably necessary, in recommending new mechanical improvements, not only to point out what alterations ought to be made, but also to show distinctly *how the work to be done can be executed in the easiest and best manner*, the fear of being by some thought prolix and tiresome must not deter me from being very particular and minute in my descriptions and instructions.

In justice it ought always to be remembered that my object in writing is professedly to be useful, and that I lay no claim to the applause of those delicate and severe judges of literary composition, who read more with a view to being pleased by fine writing than to acquire information. If those who are quick of apprehension are sometimes tempted to find fault with me for being too particular, they must remember that it is not given to all to be quick of apprehension, and that it is amiable to have patience and to be indulgent. But to proceed.

As the fire employed in heating stewpans, sauce-pans, etc., may be applied in a variety of different ways, and as the form of the utensil ought in all cases to be adapted to the form of the fire-place and to the mode of applying the heat, it is necessary, in laying down rules for the construction of stewpans and kitchen boilers, to take into consideration the construction of the fire-places in which they are to be used. But kitchen fire-places, constructed on the best principles, are susceptible of a variety of different forms.

In the spacious dwellings of the rich, where large rooms are set apart for the sole purpose of cooking, a number of separate fire-places, in large masses of brick-work constructed on the principles adopted in the kitchen of Baron de Lerchenfeld, at Munich, will be found most convenient (see page 91*); but for persons of moderate fortunes, to whom the economy of house-room is an object of importance, a less expensive arrangement may be chosen.

It is very easy (as will be shown hereafter) so to arrange the implements necessary in cooking for a moderate family, as to leave the kitchen not merely a habitable, but also a perfectly comfortable and even an elegant room. All those who have seen the kitchen in my house, at Brompton (which was fitted up principally with a view to exemplify that important fact), will not doubt the truth of this assertion.

In treating the subject I have proposed to investigate in this chapter, I shall first consider what forms will be best for saucepans and stewpans that are designed to be used in fixed fire-places, and shall then show how those should be constructed which are designed to be heated in a different manner.

### *Of the Construction of Saucepans and Stewpans for fixed Fire-places.*

The reasons have already been given why stewpans and saucepans ought always to be circular. They are indeed always made in that form; but still, as they are commonly constructed, they have a fault which renders

---

* For all such fire-places, at least for all such as are destined for heating stewpans and saucepans, I am quite sure that wood is the cheapest fuel that can be used, even here in London, where it bears so high a price. It is certainly the most cleanly and most convenient, and makes the most manageable fire. I found by an experiment, made on purpose to ascertain the fact, that any given quantity of wood, burned in a closed fire-place, gives very near three times as much heat as it would give if it were first reduced to charcoal, and then burned in the same fire-place. But the great advantage of using wood as fuel in the small fire-places of stewpans and saucepans is the facility with which it may be kindled, and the facility and quickness with which the fire may be put out (by shutting the dampers) when it is no longer wanted.

them but ill adapted for the closed fire-places I have recommended. Their handles being fastened to them on their outsides (by rivets), the regularity of their form is destroyed, and they cannot be made to fit well to the circular openings in their fire-places, which they ought to occupy and to fill.

There are two ways in which this imperfection may be remedied: the first, which is the least expensive, but which is also at the same time the least perfect, is to rivet the handle to the *inside* of the saucepan. This leaves the *outside* of the saucepan circular or cylindrical, that is to say, if care is taken to beat down the heads of the riveting nails, and to make them flat and even with the outside surface of the vessel; but the regularity of the form of the inside of the saucepan will in this case be spoiled by that part of the handle that enters the saucepan, which circumstance will not only render it more difficult to keep the saucepan clean, but will also make it impossible to close it well with a circular cover. The cover may indeed be so contrived as to fit the opening of the saucepan by making a notch in one

Fig. 23.

side of it to receive that part of the handle which is in the way; and in this manner I have sometimes caused kitchen utensils already on hand to be altered and made to serve very well for closed fire-places. The Figs. 23

and 24 will give a perfect idea of the manner in which these alterations were executed.

Fig. 24.

But, when new saucepans and stewpans are constructed, I would strongly recommend the following more simple and more advantageous contrivance.

A circular rim of iron should be provided for each saucepan with a handle belonging to it, of the form here represented; and, by forming the saucepan to this

Fig. 25.

rim, its form at its brim will be circular *within* and *without;* and consequently the saucepan will exactly fit the circular opening of its fire-place, and will at the same time be exactly fitted by its *circular cover.* No attention will in that case be necessary, in putting on the cover, to place it in any particular manner or situation; and the saucepan, not being pierced with holes for rivets, will, on that account, be less liable to leak, and will also be more durable and more easily kept clean.*

* One reason is obvious why stewpans without rivets should be more durable than those which have their handles riveted to them ; but there is another reason more occult, which requires the knowledge of a late discovery in chemistry to understand. When iron and copper, in contact with each other, are placed in a situation in which they are exposed to be frequently wetted, they act on each other very powerfully, and one of the metals will soon be destroyed by rust.

The circular iron rim above recommended should be broad and flat, from $\frac{2}{10}$ to $\frac{3}{10}$ of an inch in thickness, and from $\frac{1}{2}$ an inch to $\frac{3}{4}$ of an inch in width. Its handle, which must be welded fast to it, and must project from one side of it, may be from $1\frac{1}{4}$ inch to $1\frac{1}{2}$ in width, from 6 to 8 or 10 inches long, and of the same thickness as the circular rim where it joins it.

The under side of this flat iron rim should be made perfectly flat, in order that the saucepan, by being suspended by it in its fire-place, may so completely close the circular opening of the fire-place as to prevent the smoke from coming into the room; and also to prevent (what would be much more likely to happen) the cold air of the room from descending into the fire-place, and mixing there with the flame and smoke, and afterwards going off thus heated through the chimney into the atmosphere.

The copper saucepan or stewpan is to be fastened

---

When ships first began to be covered with copper, this fact was not known, and great inconvenience was found to arise from the rapid decay of the iron bolts in the vessels so covered. As there appeared to be no remedy for this evil, it was found necessary to substitute copper bolts for iron bolts in constructing ships intended to be coppered. These effects are now known to depend on what (from the name of its discoverer) has been called the *Galvanic influence*.

It appears to me to be highly probable that stewpans and saucepans, constructed in the manner above described, would last more than twice as long as those made in the usual manner. Frequent attempts have been made to line copper boilers and saucepans with tinned iron (commonly called sheet iron) in order to guard against the poisonous qualities of the copper; but none of these have succeeded so well as was expected, the tin being found to be destroyed by rust with uncommon rapidity. This, no doubt, was owing to the influence of the same cause by which the iron bolts of coppered ships were so suddenly destroyed.

If handles must be riveted to the sides of copper saucepans or boilers, such handles should be made of copper and not of iron; and the nails by which they are fastened should likewise be copper. They would cost something more at first, but the utensils would last so much longer that they would turn out to be much the cheapest in the end.

to its iron rim by being turned over its outward edge;
and in order that the copper, thus turned over the outward edge of the iron rim, may hold fast without projecting below the level of the lower flat surface of the
ring (which would be attended with inconvenience), the
lower part of the outward edge of the ring must be
chamfered away in the manner represented in the
following figure (26), which shows a vertical section of
the ring, of the full size, with the copper turned over it.

Fig. 26.

The upper inside edge of this iron ring may be
rounded off, as it is represented to be in the above
figure. In this figure the section of the ring is distinguished by diagonal lines, and that of the copper
(which is turned over it) by two parallel crooked lines.

When stewpans and saucepans are constructed on
the principles here recommended (with flat circular
iron rings), an advantage will be attained, which in
many cases will be found to be of no small importance: they will be well adapted for being used in small
portable fire-places heated by charcoal, or in portable
stoves heated (or rather kept hot) by heaters. Descriptions of these portable fire-places and heater-stoves will
be given in the sequel of this work.

As the upper part of the circular opening of the fire-
place (Fig. 27), on the top of which the lower part of
the circular rim of the saucepan reposes, is nearly on
a level with the top of the solid mass of the brick-work,

it is necessary that the handle of the saucepan should be bended upwards, so as to be above the level of the brim of the saucepan; otherwise, when the saucepan is in its place, there would not be room between the handle and the surface of the brick-work for the fingers to pass in taking hold of the handle to remove the saucepan. This is evident from a bare inspection of the following figure (27), which represents the section of a saucepan constructed on the plan here proposed, fitted into its fire-place.

Fig. 27.

There should be a round hole, about a ¼ of an inch in diameter, near the end of the handle, by which the saucepan may occasionally be hung up on a nail or peg when it is not in use. The cover belonging to the saucepan may be hung up on the same nail or peg, by means of the projection of its rim.

These will be thought trifling matters; but it must not be forgotten that convenience and the economy of time are often the result of attention to the arrangement of things apparently of little importance.

In constructing the cover of a saucepan, care must be taken to avoid a fault, into which it is easy to fall,

and which, as I have found by experience, will be attended with disagreeable consequences. The circular
plate of tin, or of thin sheet copper tinned, which forms
the bottom of the cover, should be of the same diameter *precisely* as the outside of the brim of the saucepan.

I once thought it would be better to make the bottom of the cover rather *larger* than the top of the brim
of the saucepan, as it is represented in the following
section : —

Fig. 28.

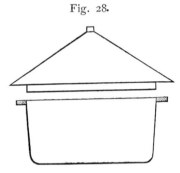

I imagined that it would prevent any thing that
happened by accident to be spilled on the cover from
finding its way into the saucepan and spoiling the victuals, and this indeed it would do most effectually ; but
it often occasioned another accident not less disagreeable in its effects. It drew the smoke into the saucepan, which happened to escape by the sides of the
circular opening of the fire-place.

When the cover is precisely of the same diameter
as the brim of the saucepan, there is little danger of
any thing entering the saucepan in this manner, as
will be evident from an inspection of the following
figure : —

Fig. 29.

The bottom of the cover may either be made quite flat, as in this section : —

Fig. 30.

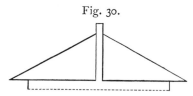

Or it may be made concave, and of a conical form, thus : —

Fig. 31.

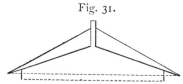

Or concave, and of a spherical figure, as is represented in the following figure : —

Fig. 32.

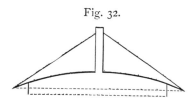

The only utility derived from making the bottom of the cover hollow instead of flat is that a little more room is left for the boiling up or swelling of the contents of the saucepan.   Cooks will be best able to judge how far this is an object of importance.

In each of the three last figures a section of the tube which carries off the steam is shown, as also a section of the rim of the cover that enters the saucepan.   This rim, which may be from $\frac{3}{4}$ of an inch to 1 inch in breadth, should be made to fit the opening of the saucepan with some degree of nicety; but it should not be fitted so closely as to require any effort in  removing it, or so as to render it necessary to use both hands in doing it, — one to hold the saucepan fast in its place, and the other to take off its cover.

The steam-tube of the cover, which may be $\frac{1}{2}$ an inch or $\frac{1}{3}$ of an inch in diameter, and should  project about $\frac{1}{2}$ an inch above the top of the cover, must pass through both the top and the bottom of the cover, and must be well fitted and soldered in both, in order that the air between the top of the cover and its bottom may be confined and completely cut off from all communication with the steam, and also with the external air.   This steam-tube should have a fit stopple, which may be made of wood, and which, to prevent its being lost, should be attached to the top of the cover by a small wire chain about 2 or 3 inches long.

In respect to the handles of these covers, the choice of the form to be adopted may be left to the workman who is employed to make the cover; for, excepting in certain cases, which will be particularly noticed hereafter, it is a point of little importance.

It is right that I should observe here that though the covers I have here described are such as I have generally recommended, yet others of different forms may be constructed on the same principles, that very possibly may answer quite as well as these, and cost less. The steam-tube, for instance, for small saucepans, may with safety be omitted, and the steam be left to make its way between the rim of the cover and the saucepan ; and, should it be thought an improvement, the upper part of the cover, instead of being a cone, may be a segment of a sphere.

The following figure is the section of the cover of a saucepan now in general use in this country. It is

Fig. 33.

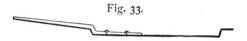

made of a circular piece of sheet copper, and its handle, which is of iron, is fastened to it by rivets ; and it is tinned on the under side. Its form is such that it fits without a rim into the saucepan to which it belongs.

This cover might be greatly improved, and perhaps rendered as well adapted for confining heat as any metal cover whatever, merely by covering it above with a thin circular plate of tinned iron or of copper, either quite flat or convex, like that represented by this figure : —

Fig. 34.

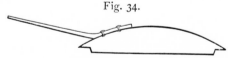

It can hardly be necessary for me to observe that this thin circular plate must be well soldered to the cover all round its circumference, in order to confine the air that is intercepted between the upper surface of the cover and the lower surface of this plate.

For the mere purpose of confining the heat in a stewpan or small boiler — were superior neatness and cleanliness not objects of particular attention — one of the very best covers that could be used would be a common saucepan cover, defended above from the cold air of the atmosphere by a circular cover of wood firmly fixed to it by means of a screw or a rivet.

The following figures represent covers so defended ; and were the circular piece of wood to prevent its

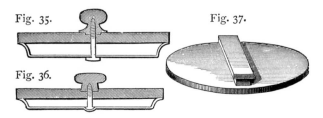

Fig. 35.    Fig. 37.

Fig. 36.

warping to be composed of two or three very thin boards, glued fast to each other and nailed or riveted together to unite them more strongly, I am inclined to think that this would be one of the best covers for common use, especially for large stewpans, that could be made.  Its handle might be made of wood, and of either of the forms represented in these figures, or of any other simple form.

The covers for large stewpans should always be furnished with steam-tubes, in order that the steam, when it becomes too strong to be confined, may escape without deranging or lifting up the cover.

A cover made entirely of wood might answer very well for confining heat, especially if care were taken to construct it in such a manner as to prevent its being liable to be warped by the heat and by the moisture to which it is continually exposed ; but the wooden covers of boilers, saucepans, and stewpans, require much attention to keep them clean, unless they be lined with tin or with sheet copper.

Having now finished my observations on the covers of small boilers and saucepans, *in their most simple state,* when they are designed merely for confining heat, it remains to consider of the means that may be put in practice to render them useful in *directing* the heat that escapes in the steam, which is formed when liquids are boiled in the various processes of cookery, and *employing this heat to useful purposes.*

As the quantity of heat that exists in steam is very considerable (as has been elsewhere observed), the recovery of this heat is frequently an object deserving of attention ; but, before we proceed in this inquiry, it will be necessary to say something respecting the method of *cooking in steam.* This subject will be treated in the following chapter.

# CHAPTER VIII.

*Of cooking in Steam. — Objections to the Steam-kitchens now in Use.—Principles on which a steam Apparatus for cooking should be constructed. — Descriptions of fixed Boilers for cooking with Steam. — A particular Description of a* STEAM-RIM *for Boilers by Means of which their Covers may be made steam-tight. — Description of a* STEAM-DISH *to be used occasionally for cooking with Steam over a Kitchen Boiler. — Account of what has been called a* FAMILY BOILER : *many of them have already been sold, and have been found very useful. — Hints to Cooks concerning the Means that may be used for improving some popular Dishes.*

A S the art of cooking with steam is well known, and has long been successfully practised in this country, it would be a waste of time to attempt to prove what is universally acknowledged ; namely, that almost every kind of food usually prepared for the table in boiling water may be as well cooked, and in many cases better, by means of boiling-hot steam. I shall therefore confine my present inquiries to the investigation of the best methods of confining and directing steam, and employing it usefully with the most simple and least expensive apparatus.

Steam-kitchens, as they are called, consist of very expensive machinery, and I have been informed, by several persons who have used them, that they do not produce any considerable saving of fuel. Bare inspection is, indeed, sufficient to show that they cannot be economical in that respect; for the surface of the tin steam-vessel filled with hot steam that is exposed quite naked to the cold air of the atmosphere is so great, that it must necessarily occasion a very considerable loss of heat.

A primary object in contriving a steam apparatus for cooking should be to prevent the loss of heat *through the sides of the containing vessels;* and this is to be done, first, by exposing as small a surface as possible to the atmosphere ; and, secondly, by covering up that surface with the warmest covering that can conveniently be used, to defend it from the cold air.

The steam-vessel in the kitchen of the Foundling Hospital is a large wooden box lined with tin, capable of containing a large quantity of potatoes; and the steam comes through a small tin tube from an oblong quadrangular iron boiler which is used daily for boiling meat, etc., for the Hospital. As this boiler is furnished with what I have called a *steam-rim* (which will presently be described), when the (wooden) cover of the boiler is down, all the steam that is generated in the boiler is forced to pass through the steam-box, and the potatoes, greens, etc., that are in the box are cooked without any additional expense of fuel.

The steam-box has a steam-rim and also a wooden cover which, when it is down, closes the box and makes it perfectly steam-tight.

When steam is generated faster than it can be condensed in the steam-box, that which is redundant passes off by a waste-tube, which conducts it into a neighbouring chimney.

The apparatus for cooking with steam in the kitchen of the House of Correction, at Munich, is still more simple. Here two equal quadrangular boilers are set, one at the end of the other, at the same level, in the same mass of brick-work; and the flame and smoke from the same fire pass under them both (see Plate X., Fig. 7, and Plate XI., Fig. 9). Both boilers being enclosed in brick-work and being covered with wooden covers, it is evident that no part of the apparatus is exposed to the cold air. I say *no part* of it; for the covers of the boilers being of wood, which is one of the worst conductors of heat, very little heat can make its way through them; and to prevent even this loss, inconsiderable as it is, these wooden covers may, if it should be thought necessary, be defended from the cold air by warm rugs thrown over them.

The smoke which passes under the second boiler not only prevents the approach of the cold air to the under surface of its bottom, but, acting on the small quantity of water that is contained in it, actually assists in the generation of steam. It even happens sometimes (namely, when there is but a small quantity of water in the second boiler, and the first is nearly filled with cold water) that the water in the second boiler actually boils and fills the boiler with steam, before the water in the first boiler is heated boiling-hot.

This appears to me to be one of the most economical methods that can be used for cooking, and that it is well adapted for hospitals and also for large private families. If it should be necessary to make provision for cooking a great number of different dishes in steam at the same time, either the steam-boiler may be made sufficiently large to receive them, or, instead of it, two or more steam-boilers of a moderate size may be put up; and, if the different kinds of food that are cooked at the same time in the same steam-boiler be placed each in a separate dish and covered over with some proper vessel in the form of a bell (a common earthen pot, for instance, turned upside down), the exhalations from the different kinds of food will be prevented from so mixing together as to give an improper taste or flavour to any of the victuals.

These covers to the different dishes will likewise be useful on another account. When the cover of the steam-boiler is opened for the purpose of examining or of introducing or removing any dish, the process of cooking going on in the other dishes will not be interrupted, for their bell-like covers, remaining filled with steam, will prevent the cold air from coming into contact with the victuals. It is true that the cover or lid of the steam-boiler must not be kept open too long, otherwise the steam confined under the covers of the dishes will be condensed, and the cold air will find its way under them.

In order that these boilers may be perfectly steam-tight when their lids are down, they must all be furnished with *steam-rims;* and there must be a tube of communication between them for the passage of the steam, and another tube to carry off the redundant steam from the boiler which is situated farthest from the fire.

If it should be necessary, the principal boiler may, without any difficulty or inconvenience, be divided into two compartments, so as to render it possible to prepare two different kinds of soup, or to boil two different things separately at the same time. Suppose, for instance, that the apparatus is designed for the kitchen of a large family, and that the principal boiler is 12 inches wide, 24 inches long, and 12 inches deep. This may be so divided by a vertical partition as to form two compartments: the one, that immediately over the fire, for instance, 12 inches by 10; and the other, 12 inches by 14. In this case I should make the second, or *steam-boiler*, 24 inches square by 12 inches deep, and should cause the smoke to circulate in three flues parallel to each other. The first (in the hither end of which the fire-place should be situated) should be immediately under the first boiler, and the second and third should be under the second boiler.

The following figure shows the manner in which these boilers should be set : —

Fig. 38.

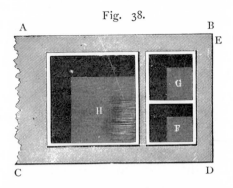

A, B, is the side of the room; A, C, D, E, the mass of brick-work in which the boilers are set; F and G are the two compartments of the first boiler, which is shown with its steam-rim; H is the larger boiler, which is also represented with its steam-rim.

The covers of these boilers (which do not appear in the figure) should be so attached to the boilers by hinges as to be laid back when the boilers are opened, and rested against the side of the room; and these covers should be lined with tin or with thin sheet copper tinned.

Fig. 39.

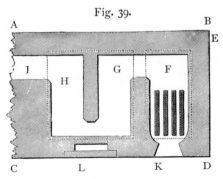

The foregoing figure represents a horizontal section of the brick-work in which these boilers are to be set, taken at the level of the tops of the flues.

A, B, is the side of the room; and A, C, D, E, the mass of brick-work which is placed against it; F, G, and H are the three parallel flues; and I is the canal that carries off the smoke from the second boiler to the chimney; K is the opening into the fire-place by which the fuel is introduced; and L is a passage, closed up with a tile or with loose bricks, which is occasionally opened to clean the flues, G and H. The damper in the canal, I, may be placed near the left-hand side of the second boiler. The situations of the boilers are indicated by dotted lines.

As it is not necessary that I should repeat in this place the directions which have already been so amply explained concerning the proper method of proceeding in setting boilers, I shall not enlarge farther on that subject, but shall proceed to give an account of a very essential part, not yet described, of the apparatus necessary for cooking with steam in the simple way I have here recommended: the part I mean is the *steam-rim* of the boiler.

### *Description of a Steam-rim for a Boiler, by Means of which its Cover may easily be made steam-tight.*

To give a more complete idea of this contrivance, I have, in the following figure, represented a vertical section of a small part of one side of a boiler and its steam-rim with its (wooden) cover in its place, both of one half size.

A, B, is a section of part of the flat wooden cover; the crooked line, C, D, is a section of the steam-rim, and part of the side of a boiler; E is a section of a descending rim of wood belonging to and making an essential part of the cover, which rim, when the cover is down, enters the steam-rim of the boiler, and reposes on the bottom of it. In the figure it is represented in this situation: the wooden rim of the cover is fastened

Fig. 40.

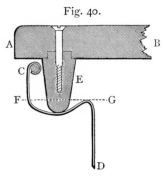

to the flat part of it by means of wood-screws, one of which is represented in the figure.*

Now it is evident, from an inspection of the figure, that a small quantity of water will lodge in the steam-rim, and will stand at the level of the dotted line, F, G; and, as the rim of the cover will enter this water when the cover is shut down, all communication between the steam in the boiler and the external air must necessarily be cut off, and of course the steam will be completely confined.

It is true that, if in consequence of the increase of its temperature above the heat of water boiling in the open air the elasticity of the steam should become sufficient to overcome the pressure of the atmosphere,

---

* The cover itself is supposed to be framed and panelled in the manner described in the fifth chapter of this Essay, and it should be lined with tin or with thin sheet copper tinned, in order to prevent the wood from being cracked and destroyed by the steam.

it will force the water in the steam-rim to ascend toward C, and, getting under the rim, E, of the cover of the boiler, it will make its escape, but no bad consequences will result from this loss; on the contrary, the steam-rim will in this case serve instead of a *safety-valve*. And, although this contrivance may not be adequate to the confining of *strong* steam, it certainly answers perfectly well for confining that kind of steam which is most proper to be used for cooking. It will likewise be found useful in many cases for covering boilers, where the principal object in view is to prevent the contact of the cold air with the contents of the boiler. It will be useful for the boilers of bleachers, as also for laundry boilers, for brewers' boilers, and for all boilers destined for the evaporation of liquids under a boiling heat.

It appears to me that this contrivance might, with a little alteration, be used with great advantage for covering the boilers used by distillers. By making the steam-rim deeper, the cover of the boiler would be tight, under a considerable pressure; and by making the boiler broad and shallow, with several separate fire-places under it (the flat bottom of the boiler being supported on the tops of the flues of these fire-places), a variety of important advantages would be gained, and these would not be compensated by any disadvantages that I can foresee. The boiler might be constructed of very thin sheet copper, which would not only render it less expensive, but would also make it more durable.

When steam-rims were first introduced, they were made of the form represented in the following figure, which represents a vertical section of part of one side of a boiler with a steam-rim, covered with a conical double cover made of tin: —

Fig. 41.

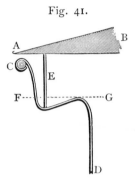

In this and the following figures, A, B, represents a section of part of one side of the (double) cover of the boiler; C, D, the steam-rim and part of one side of the boiler; E, the descending rim of the cover; and F, G, the level of the water in the steam-rim, — all of one half size.

This construction was found to be attended with an inconvenience, which, indeed, might easily have been foreseen. When the steam, on being confined, became strong enough to force its way under the descending rim, E, of the cover of the boiler, the water in the steam-rim was frequently blown out of it with considerable violence and dispersed about the room. To prevent these disagreeable accidents, the form of the upper part of the steam-rim was altered. To make a proper finish to the boiler, the edge of its brim (which forms the top of its steam-rim) had been turned *outwards* over a strong wire. It was now turned *inwards* over the wire; and the outside or rising part of the steam-rim, instead of being made *sloping outwards*, was now made *vertical*.

A complete idea of these different alterations, and of

the effects necessarily produced by them, may be formed by comparing the foregoing figure (No. 41) with the following : —

Fig. 42.

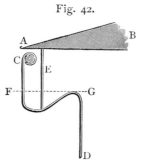

It is evident that in this case, as there is sufficient room between the outside of the descending rim of the cover and the vertical side of the steam-rim to contain all the water that can be forced upwards between them by the steam, there is little danger of any part of this water being blown out of the steam-rim by the steam when it makes its escape under the rim of the cover.

*Of the Manner in which Kitchen Boilers and Stewpans may be constructed so as to be rendered useful in cooking with Steam.*

If a common kitchen boiler be furnished with a steam-rim, and the descending rim of its cover be made to shut down into it, the steam in the boiler will be effectually confined, and may in various ways be usefully employed in cooking. One of the simplest methods of doing this is to set what I shall call a *steam-dish* upon the boiler. The bottom of this steam-dish being furnished with a descending rim or projection, fitting into the steam-rim of the boiler, the steam-dish may be made to serve as a cover to the boiler; and, if a number of small holes be made in the bottom of this dish near its circumference, the steam will pass up into it from below; and, if it be properly closed above, any victuals placed in it will be cooked in steam.

If this dish be furnished with a steam-rim of the same form and size with that of the boiler, the cover of the boiler will then serve for covering the steam-dish, whenever that dish is in use.

The following figure, which represents a vertical section of the apparatus, will show this contrivance in a clear and distinct manner: —

Fig. 43.

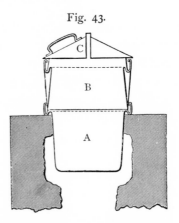

A is the boiler, which is seen set in brick-work; B is the steam-dish; and C is the cover of the boiler, which is here made to serve as a cover for the steam-dish.

The sides of the steam-dish (which is made of tin) are double, for the purpose of confining the heat more effectually.

If it be required to cook several kinds of food at the same time, a steam-dish may be used that is divided into several compartments; or two or more steam-dishes may be placed one above another over the same boiler, that which is uppermost being covered with the cover of the boiler.

A very complete apparatus of this kind may be seen in the kitchen of Mr. Summers, of New Bond Street, ironmonger, who makes and sells these articles, and who has sold no less than 225 sets of these family boilers, as they are called, since he first began to manufacture them; and Mr. Feetham, of Oxford Street, has sold 110 sets of them. A cooking apparatus of this kind may likewise be seen at the Royal Institution; and at Heriot's Hospital, at Edinburgh; and in the houses of many private families ih England and Scotland. There are several tradesmen who now manufacture them; and all persons desirous of making and selling them are at full liberty to do so.

When different kinds of food, placed one above the other, are cooked in steam, the drippings of those above might, in some cases, be apt to spoil those below if means were not used to prevent it. This inconvenience may be avoided in the apparatus I am describing by introducing the food into the steam-dishes, placed in deep plates or in shallow basins, sufficiently capacious, however, to contain as much water as will be generated in consequence of the condensation of the steam on the surface of the food in heating it boiling-hot. I say "in heating it boiling-hot;" for, after it is once heated to that temperature, no more steam will be condensed upon it, however long the process of cooking may be continued.*

---

* It is not difficult to determine with great precision what the size or contents of the dish must be, in order that it may contain all the water that can possibly be produced by the condensation of the steam, in heating the victuals

This is a curious circumstance, and the knowledge of the fact may be turned to a good account. If, for instance, it were required to make the strongest extract of the pure juices of any kind of meat, unmixed with water, this may be done by heating the meat nearly boiling-hot, either in boiling water or in steam, and then putting it, placed in a shallow dish, into a steam-dish, or into any closed vessel filled with hot steam, and leaving it in this situation two or three hours, or for a longer time. Whatever liquid is found collected in the dish at the end of the process must necessarily be the purest juices of the meat. In this manner the richest gravies may no doubt be prepared.

that are cooked in it to the temperature of boiling water. Suppose, for instance, that a piece of beef weighing six pounds is to be cooked in the steam-dish, and that this meat, when it is put into the dish, is at the temperature of 55° of Fahrenheit's thermometer, which is the mean annual temperature of the atmosphere at London. Now as this piece of meat is to be made boiling-hot, its temperature must be raised 157 degrees, namely, from 55° to 212°. But we have seen that any given quantity, by weight, of beef, requires less heat to heat it any given number of degrees, than an equal weight of water, in the proportion of 74 to 100 (see the introduction to this Essay, page 183); consequently these 6 lbs. of beef will be heated 157 degrees, or from 55° to the boiling point, with a quantity of heat which would be required to heat 4 lbs. 7 oz. of water 157 degrees.

Now if we suppose, with Mr. Watt, that the steam which produces, in its condensation, 1 lb. of water gives off as much heat as would raise the temperature of 5¼ lbs. of water 180 degrees, namely, from the point of freezing to that of boiling water, the same quantity of heat must be sufficient to raise the temperature of 6 lbs. 5 oz. of water 157 degrees, or from 55° to 212°.

And if 6 lbs. 5 oz. of water require 1 lb. of condensed steam to heat it 157 degrees, 4 lbs. 7 oz. of water, or 6 lbs. of beef, will require only 11¼ oz. of condensed steam to raise its temperature the same number of degrees, for it is 6 lbs. 5 oz. is to 1 lb. as 4 lbs. 7 oz. to 11¼ oz.

Consequently, if 6 lbs. of beef at the temperature of 55° were placed in a steam apparatus, in a shallow dish capable of containing 11¼ oz., or a little less than *three quarters of a pint*, this dish would contain all the water that could possibly result from the condensation of steam on the surface of the meat, in heating it boiling-hot.

This computation may be of some use in determining the dimensions of the vessels proper to be used for holding the victuals that are cooked in the steam-dishes above described.

Thick steaks or cutlets of beef, boiled in this manner, and made perfectly tender throughout, and then broiled on a gridiron, and served up in their own gravy, with or without additions, would, I imagine, be an excellent dish, and very wholesome. But it must be left to cooks and to professed judges of good eating to determine whether these hints (which are thrown out with all becoming humility and deference) are deserving of attention. For, although I have written a whole chapter on the pleasure of eating, I must acknowledge, what all my acquaintances will certify, that few persons are less attached to the pleasures of the table than myself. If, in treating the subject, I sometimes appear to do it *con amore*, this warmth of expression ought, in justice, to be ascribed solely to the sense I entertain of its infinite importance to the health, happiness, and innocent enjoyments of mankind.

## CHAPTER IX.

*Description of a* UNIVERSAL KITCHEN BOILER, *for the Use of a small Family, to answer all the Purposes of Cookery; and also for boiling Water for Washing, etc. — Description of a* PORTABLE FIRE-PLACE *for a universal Kitchen Boiler. — Account of a Contrivance for warming a Room by Means of this Fireplace and Boiler. — Of* STEAM STOVES *for warming Rooms. — They are probably the best Contrivance for that Purpose that can be made Use of, — they warm the Air without spoiling it, they economize Fuel, and may be made very ornamental.*

*Description of a* UNIVERSAL KITCHEN BOILER *for the
Use of small Families, to answer all the Purposes of
Cookery; and also for boiling Water for Washing, etc.*

THE following figure represents a vertical section
of this boiler, and also of its fire-place and cover.

This boiler is supposed to be made of cast iron, and
its section is represented by a double line.   The lower

Fig. 44.

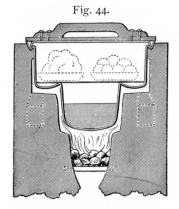

part of it, which is represented as being filled about
half full with water, is 12 inches in diameter above,
about 11 inches in diameter below, and 9½ inches deep.
The upper part of it, which is furnished with a steam-
rim, is 24 inches in diameter above — where its steam-
rim begins — and 23 inches in diameter below — where
it joins the flat part which unites it to the lower part
of the boiler.

The lower part of this boiler (which might, without
any impropriety, be called the *lower boiler*) is destined
for containing the soup or the water that is made to
boil, while the upper and broader part is used for boiling with steam.   The brim of the lower boiler projects
upward, about an inch above the level of the flat bot-

tom of the upper boiler. This projection prevents the water resulting from the condensation of steam against the sides of the upper boiler from descending into the lower boiler. The upper boiler is $8\frac{1}{2}$ inches deep, from the top of the inside of its steam-rim to the flat part of its bottom. The whole depth of both boilers is 18 inches, from the top of the steam-rim to the lower boiler.

A circular piece of tin, about 22 inches in diameter, with many holes through it to give a free passage to the steam, being laid down in a horizontal position upon the top or projecting brim of the lower boiler, upon this circular plate the shallow dishes are placed, which contain the victuals that are to be cooked in steam. Two such dishes are faintly represented in the foregoing figure by dotted lines.

The cover of this universal boiler is a shallow circular dish, 26 inches in diameter at its brim, and about $1\frac{1}{2}$ inches deep, turned upside down, and covered above with a circular covering of wood to confine the heat. The handle to this cover is a strong cleat of wood, fastened to the circular wooden cover by means of four wood screws. This handle is distinctly represented in the figure.

The circular wooden cover for confining the heat must be constructed in panels, and must be fastened to the shallow metallic dish by means of rivets or wood screws. In doing this, all the precautions must be taken that are pointed out in the fifth chapter of this Essay, page 185; otherwise the wood and the metal will be separated from each other, in consequence of the shrinking of the wood on its being exposed to heat.

The inverted shallow dish, which, properly speaking,

constitutes the cover of this boiler, may be made either of tin or of sheet iron or of sheet copper; or it may be made of cast iron. Whatever the material is of which it is constructed, care must be taken to make it of such dimensions precisely that its brim may enter the steam-rim, and occupy the lower or deepest part of it, otherwise the steam will not be properly confined in the boiler.

The following figure represents a vertical section, *of one half size*, of the steam-rim of one of these boilers (of cast iron), together with a section of a part of an inverted shallow cast iron pan, which serves as a cover to the boiler, and also of the circular covering of wood which is attached to the pan, and defends it from the cold air of the atmosphere.

Fig. 45.

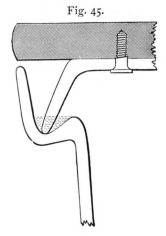

In this figure the steam-rim is represented as being full of water, and one of the screws is seen which fasten the circular wooden cover to the inverted shallow pan which confines the steam in the boiler.

On examining the two preceding figures, it will be

found that both the boiler and its cover are of forms that will readily deliver from their moulds; and that circumstance will enable iron-founders to sell these articles at low prices.

The mass of brick-work in which this boiler is set may be a cube of 3 feet; or, by sinking the ash-pit in the ground, its height may be reduced to $2\frac{1}{2}$ feet.

In order that the flame may be made to separate and spread equally on all sides under the lower boiler, the smoke should be made to pass off in two small canals situated on opposite sides of the boiler. The openings of these canals may be a little below the level of the bottom of what has been called the upper boiler; and the smoke, being made first to descend nearly to the level of the bottom of the lower boiler, may then pass off horizontally towards the chimney. The situation of the two horizontal canals (on opposite sides of the boiler) by which the smoke goes off is indicated (in Fig. 44) by dotted lines.

So much has already been said in the foregoing chapters relative to the construction of closed fireplaces for kitchen boilers, that it would be quite superfluous to give any particular directions respecting the construction of the fire-place for this boiler. The manner in which the boiler is set in brick-work, and the means that are used for causing the smoke to surround it on every side, are distinctly shown in the figure.

In order more effectually to confine the heat, the boiler should be entirely enclosed in the brick-work on every side, in such a manner that the brim of its steam-rim should not project above it more than half an inch. To preserve the brick-work from being wetted, the top of it may be covered with sheet lead, which may be made to turn over the top of the brim of the steam-rim of the boiler.

There may either be a steam-tube in the cover of the boiler, or the steam may be permitted to force its way under the descending rim of the inverted shallow pan which constitutes the cover. If there be a steam-tube, it should be half an inch in diameter and about one inch in length; and it should be made very smooth on the inside, in order that another tube of tin or of tinned copper, about 10 inches in length, may pass freely in it. The use of this movable tube is to cause the air to be expelled from the upper boiler, while it is used for cooking with steam. This will be done if, while the water below is boiling, the long tube be thrust down into the boiler through the steam-tube till its lower end comes to the level of the brim of the lower boiler. For, as steam is considerably lighter than common air, it will of course rise up and occupy the upper part of the upper boiler, and the air below it being compressed will escape through the tube we have just described; and, although that tube should remain open, the upper boiler will nevertheless remain filled with steam, to the total exclusion of atmospheric air. The inside of the steam-tube and the outside of the movable tube should be made to fit each other with accuracy, in order that no steam may escape between them. The necessity of this precaution is too evident to require any elucidation.

It will be best to place the steam-tube within about an inch of the side of the cover, in which case it will be easy, by turning the cover about, to place it in such a position that the movable tube may descend into the upper boiler without being stopped by meeting with any of the dishes that are placed in it.

It is hardly necessary that I should observe here that boilers on the principles above described may be constructed of sheet iron or sheet copper as well as of cast iron, and that they may be made of any dimensions. That which is represented in the foregoing figure (No. 44) is of a moderate size, and would, I should imagine, be suitable for the family of a labourer consisting of eight or ten persons. The lower part of the boiler would hold about $3\frac{3}{16}$ gallons; but the whole boiler, filled up to within an inch of the level of the inside of the steam-rim, would hold $14\frac{1}{4}$ gallons. When so filled up, I should suppose the boiler to be sufficiently capacious to heat water for washing or for any other purpose that could be wanted by an industrious family consisting of the number of persons above-mentioned.

*Description of a* PORTABLE FIRE-PLACE *for a* UNIVERSAL KITCHEN BOILER.

The following figure represents a vertical section of the fire-place with its boiler in its place : —

Fig. 46.

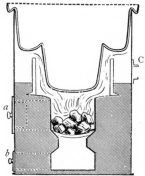

This figure is drawn to a scale of 20 inches to the inch.

The boiler is supposed to be of cast iron, and the section of it is represented by a double line. To render its form more conspicuous, its cover is omitted.

The portable fire-place is a cylinder of sheet iron, $24\frac{3}{4}$ inches in diameter, and $34\frac{1}{4}$ in height, open above and closed below. The sections of this cylinder and of its bottom are marked by strong black lines.

The fire-place, properly so called, is the centre or axis of this cylinder. It is built of fire-bricks and Stourbridge clay, and the fire burns on a circular cast iron dishing-grate, 8 inches in diameter.

The opening (at *a*) by which the fuel is introduced is marked by dotted lines, as is also another opening below it (at *b*) which leads to the ash-pit. These openings are closed by doors of sheet iron, which are attached by hinges to the outside of the cylinder, and fastened by means of turn-buckles.

The door of the ash-pit is furnished with a register for regulating the admission of air.

The smoke is carried off by a horizontal tube, a part of which is seen at C.

There is a particular and very simple contrivance for causing the smoke to come into contact with the sides of the lower boiler and with the flat bottom of the upper boiler, and then to *descend* before it is permitted to pass off. This is a cylinder of cast iron or of earthenware, which is 16 inches in diameter within or in the clear, and 8 inches high, with a thin flange about an inch wide at its lower extremity. This flange serves as a foot for keeping it steady in its vertical position, and also for fastening it in its place by laying the ends of a circular row of short pieces of brick upon it. The lower end of this cylinder being set down at the level of the bottom of the lower boiler, upon the top of the hollow cylindrical mass of brick-work which constitutes the fire-place, the smoke is obliged to pass up between the inside of this cylinder and the outside of the lower boiler and to strike against the flat bottom of the upper boiler. It then passes horizontally over the top of this cylinder, and, turning downwards into the space which is left for it between the outside of this short cylinder and the great cylinder of sheet iron in which the boiler is suspended, it passes off by the small horizontal tube which carries it to the chimney.

This short cylinder is so distinctly represented in the figure that letters of reference are quite unnecessary.

A piece of brick or of fire-stone, about $2\frac{1}{2}$ inches thick, is supposed to be attached to the inside of the fire-place door, to prevent its being too much heated by the fire; and this is represented in the figure by dotted lines. The knobs in the fire-place door and in the door of the ash-pit are designed to be used as a handle in opening them.

This portable fire-place may have two strong handles for transporting it from place to place; and, as the boiler may be removed and carried separately, the fire-place will not be too heavy to be carried very conveniently by two men.

Without stopping to expatiate on the usefulness of this new implement of cookery, I shall proceed to show how its utility may be made still more extensive. With a trifling additional expense it may be changed into one of the very best stoves for warming a room in cold weather that can be contrived. I say one of the *very best*, for it will warm the air of the room without its being possible for it ever to heat it so much as to make it unwholesome; and it will do it with the least trouble and at the expense of the least possible quantity of fuel.

*Description of a Contrivance for warming a Room by Means of a portable universal Kitchen Boiler.*

The following figure represents an elevation, or front view, of the machinery that may be used for this purpose : —

Fig. 47.

This machinery is very simple. It consists of the portable boiler and fire-place represented in the preceding figure (No. 46), with an inverted cylindrical vessel, constructed of tin or of very thin sheet copper, placed over the boiler. This cylindrical vessel, which I shall call a steam-stove, must be just equal in diameter to the steam-rim of the boiler at the lowest or deepest part of that rim; and it may be made higher or lower, according to the size of the room that is to be heated by it. That represented in the foregoing figure is 26 inches in diameter and 24 inches high, which gives 17 square feet of surface for heating the room.

This *steam-stove* may be made of common sheet iron; but in that case it should be japanned within and without, to prevent its rusting. In japanning it, it might be painted or gilded, and rendered very ornamental. The portable fire-place might likewise be japanned and ornamented; but in that case it would be necessary to line that part of it with clay or cement with which the smoke comes into contact, otherwise the heat in that part might injure the japan.

There must be a small tube about $\frac{1}{4}$ of an inch in diameter in one side of the steam-stove, just above the top of the steam-rim of the boiler. This tube should be about 2 inches in length, and it should project inwards, horizontally, into the cavity of the steam-stove. Into this tube one end of another longer tube should be introduced, which is designed to carry off the redundant steam into the chimney.

The reason why this tube should be placed near the bottom of the steam-stove will be evident to those who recollect that steam is lighter than air. Were it placed at the top of it, no steam would remain in the stove, and the object of the contrivance would be defeated.

This small steam-tube at the lower part of the stove may, with safety, be kept quite open; for, unless the water in the boiler be made to boil with vehemence, little or no steam will issue out of it; for the greater part, if not the whole of it, will be condensed against the top and sides of the steam-stove.

As the water which results from this condensation of steam will all return into the boiler, it will seldom be necessary to replenish the boiler with water.

When cooking is going on in the boiler in cold weather, the steam-stove will supply the place of a cover for the boiler; but, when the weather is warm, the cover of the boiler may be used instead of it, and the air of the room will be very little heated.

Steam-stoves on these principles would be found very useful in heating halls and passages, and I think they might be used with advantage for heating elegant apartments. They are susceptible of a variety of beautiful forms, and are not liable to any objections that I am aware of. A most elegant steam-stove might be made in the form of a Doric temple, of eight or ten columns, standing on a pedestal. The fire-place might be situated in the pedestal, and the columns and dome of the temple might be of brass or bronze, and made hollow to admit the steam. In the centre of the temple a small statue might be placed as an ornamental decoration; or an Argand's lamp might be placed there to light the room. In case a lamp should be placed in the centre of the temple, there should be a circular opening left in the top of the dome for the passage of the smoke of the lamp.

The fire under the boiler may be lighted and fed without the room or within it; or the steam may be brought from a distance in a leaden pipe or copper tube. If the boiler that supplies the steam is situated in the pedestal of the temple, and if the fire is lighted from within the room, the fire-place and ash-pit doors may be masked by tablets and inscriptions.

But I need not enlarge on the means that may be

used for rendering a useful mechanical contrivance ornamental and expensive; for many persons will be ready to lend their assistance in that undertaking.

Those who wish to see one of these universal kitchen boilers will find one set in brick-work in the kitchen of the Royal Institution. It is constructed of copper, and tinned on the inside; and it is considerably larger than that I have here described. The method used for confining the steam in this boiler is different from that here recommended, and there is a contrivance for heating the contents of the boiler occasionally by means of steam, which is brought from another boiler; but this contrivance has no particular connection with the invention in question, and is introduced here merely to show how steam may be employed for making liquids boil.

In order that these universal kitchen boilers, with steam-stoves, may the more easily find their way into common use in this country, some method should be contrived for making tea in them. Now I think this might be done by putting the tea with cold water into a shallow tin tea-pot, or rather kettle, and placing it in the upper boiler, directly over the lower boiler. I once made an experiment of this kind; and, if I was not much mistaken, the tea that was so made was uncommonly good and high-flavoured. It certainly appeared to be considerably stronger than it would have been, if, with the same quantities of tea and of water, it had been made in the common way.

Boiling water poured upon a vegetable substance does not always extract from it all that might be extracted by putting the substance to cold water and heating them together. This fact is well known; and it renders it probable that the method here proposed of making tea would be advantageous. If this should be the case, no implement could be better contrived for that purpose than our universal kitchen boiler.

---

## CHAPTER X.

*Description of a new-invented* REGISTER-STOVE *or* FUR-
NACE *for heating Kitchen Boilers, Stewpans, etc.
— Of the Construction of Boilers and Stewpans
peculiarly adapted to those Stoves. — Particular
Method of constructing Stewpans and Saucepans
of Tin, by which they may be rendered very durable.
— Description of a small* PORTABLE FIRE-PLACE *for
Stewpans and Saucepans. — Of cast-iron* HEATERS
*for heating Kitchen Utensils.*

HAVING learned, by frequenting kitchens while the various processes of cookery were going on in them, how very desirable it would be that the cook might be enabled to regulate and occasionally to moderate the fires by which stewpans and saucepans are heated, I set about contriving a fire-place for that purpose, which on trial was found to answer very well. The first fire-place of this kind that was constructed was put up in my own kitchen, at Munich, where it was in daily use for more than twelve months; and soon after I returned to this country (in the year 1798) one of them was put up in the kitchen of Mr. Summers, ironmonger, No. 98 New Bond Street, where it

has been exhibited to the view of those who frequent
his shop. Since that time a great number of them
have been put up in the kitchens of private families,
and, as I am informed, are much liked. As their usefulness appears to me to have been sufficiently ascertained by experience to authorize me to recommend
them to the public, I shall now lay before the reader the
most exact and particular description of them that I can
give; premising, however, that it will be difficult to give
so clear an account of this contrivance as to enable
a person to form a perfect idea of it without having
seen it.

I shall perhaps be most likely to succeed in this
attempt, if I begin by exhibiting a view of the thing
to be described.

Fig. 48.

This plate represents a view of a register-stove fire-
place for two stewpans, actually existing in Heriot's
Hospital, at Edinburgh. It is placed in a mass of
brick-work, 2 feet 6 inches high, 4 feet 6 inches long,

and 2 feet wide from front to back, situated in a corner
of the room on the right-hand side of the fire-place.
In the middle of the front of this mass of brick-work
are seen the front of the fire-place door (which is
double), and the ash-pit register-door; and near the
end of it, on the left, in the upper front corner, may
be discovered the stone stopper, which closes a canal,
which is occasionally opened for cleaning out the soot
from the flues in the interior parts of the mass of brick-
work.   A like stopper, and which serves for a like purpose, may be seen at the end of the mass of brick-work,
near the right-hand corner above.   Each of these
stoppers is furnished with an iron ring, fastened by a
staple, which serves as a handle in removing and
replacing it.

On the top of this mass of brick-work there is laid
a horizontal plate of cast iron, 18 inches wide, 3 feet
long, and about $\frac{1}{3}$ of an inch in thickness; and on the
right and left of this iron plate, and level with its upper
surface, there are placed two flat stones, each 9 inches
wide and 18 inches long, being just as long as the
iron plate is wide.

At the back of this iron plate runs a flue, 4 inches
wide and 5 inches deep, which is covered above, at the
level of the upper surface of the iron plate, with a flat
stone, 6 inches wide.

One of the most essential parts of this contrivance is
the iron plate, with its circular register, both which are
represented by the following figure; but only one half
of the plate is represented, being shown broken off in
the middle.

In this figure the circular movable register (which is
distinguished from the oblong plate to which it belongs

by marking the latter by fine horizontal lines) is shown
in its place; and the projecting piece of metal is also
seen which serves as a handle to turn it about on its
centre.     This circular register has a shallow circular
groove near its circumference, about $\frac{1}{2}$ an inch deep and

Fig. 49.

$1\frac{1}{4}$ inches wide; and between the inside of this groove
and the centre of the register there are two holes or
openings on opposite sides of the centre which answer
to two other openings of like form and dimensions,
which are in each half of the oblong plate to which the
registers belong.     By one of these openings (that next
the middle of the oblong plate) flame rises from a fire
situated below, and spreads under the bottom of a boiler
which is suspended over the circular register; and by
the other it descends, and, again entering the mass of
brick-work, it goes off by a horizontal canal which communicates with the chimney.

The boiler or stewpan is suspended over the circular register-plate, and the heat is confined about it by
means of a hollow cylinder of sheet iron or of earthen-
ware (about one inch longer or higher than the boiler
is deep), and open at both ends, the lower end of which,

entering the shallow groove of the register, reposes on it, while its upper end is closed by the boiler which, resting on it by its brim, is suspended in it, and consequently is surrounded by the flame.

This cylinder must be made quite flat or even at its two ends by grinding it on a flat stone, and the boiler must be made to fit it accurately, not however by fitting too nicely into its opening (which method would not be advisable), but by making the under part of the iron ring which forms the projecting brim of the boiler perfectly flat, and causing the boiler to be suspended by that ring on the flat end of the cylinder.

To prevent the escape of the flame under the bottom of the cylinder or between its lower end and the circular register-plate on which it stands, a small quantity of sand or (what will be still better) of fine filings of iron or brass may be put into the groove in which the cylinder is placed; and the same means may be used for making the joinings tight between the circular registers and the flat plate to which they belong.

The following figure, which shows a vertical section of this register-stove with its fire-place and its two boilers, or rather stewpans, will give a clear idea of the arrangement of the machinery.

These stewpans, which are $10\frac{1}{2}$ inches in diameter above and 6 inches deep each, are constructed according to the directions given in the seventh chapter of this Essay. They are of copper, tinned, and are turned over flat iron rings at their brims. Their handles are not seen in this figure. Their covers, which are of tin and made double, are on a peculiar construction. They are so contrived that a small saucepan for melt-

ing butter or warming gravy may be placed upon then.
and heated by the steam from their stewpans.

From a careful inspection of the three foregoing
figures, and a comparison of them with the short description that has been given of the various parts of
this machinery, it will, I fancy, be possible to form so
distinct an idea of this contrivance as to enable any
person conversant in matters of this kind to imitate

Fig. 50.

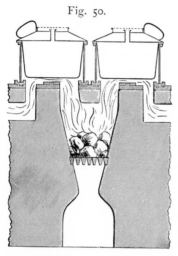

the invention, even without ever having seen the work
executed. The principles at least on which this contrivance is founded will be perfectly evident; and, when
they are understood, ingenious men will find little difficulty in the application of them to practice. It is
indeed highly probable that simpler and better means
of applying them will be found than those I have
adopted, when the use of the contrivance shall become
more general. I am indeed aware of several alterations
of the machinery which I think would be improvements; but, as I have not tried them, I dare not re-

commend them as I recommend things which I know from experience to be useful.

I shall now proceed to give an account of several precautions in the construction and use of these register-stoves for boilers, which have been found to be necessary and useful.

The circular registers are so constructed that, by turning them round, they may be so placed as either to close entirely the holes in the flat plate on which they lie, or to leave them open more or less. Now, as there is no passage by which the smoke can go off from the fire-place into the chimney but through these holes, care must be taken never to attempt to kindle the fire when both these registers are closed, and never to open one of them without having first placed a hollow cylinder on it and a fit saucepan or boiler in the cylinder, to close it above. It can hardly be necessary that I should add that care must always be taken to put water or some other liquid into the boiler to prevent its being burned and spoiled by the heat.

The state of the register, in regard to its being more or less open, cannot be seen when the boiler is in its place, as the openings of the register are concealed by it and by the cylinder in which it is suspended. But, although the state of the register under these circumstances is not seen, it is nevertheless known; and the heat which depends on the dimensions of the opening left for the passage of the flame may at any time be regulated with the utmost certainty. By means of a projecting pin or short stub, represented in the Fig. 49, belonging to the lower (fixed) plate, and which is cast with it, the movable circular register is stopped in two different positions, in one of which the openings for the flame are as wide as possible, and in the other they are quite closed. When the handle by which the circular plate is turned round is pulled as far forward as possible towards the front of the brickwork, the register is wide open. In this situation it is represented in the Fig. 49. When it is pushed as far backwards as possible, the register is closed; and its situation at any intermediate station of the handle between these two limits of its motion will at any time show the exact state of the register.

That the handles of the register plates may not interfere with each other, they are placed on the sides of their plates which are farthest from the fire; consequently they are as far from each other as possible. The form of these handles is such that they never become very hot, although they are of iron and of a piece with their plates, being cast together. The cold air of the atmosphere passing freely upward through a conical hole (left in casting) in the centre of the knob of the handle, the heat is carried off by this current of air almost as fast as it arrives from the circular plate.

There is a circumstance to which it is absolutely necessary to pay attention in setting the large flat iron plate in the brick-work, otherwise the machinery will be liable to be soon deranged by the effects of the expansion of the metal by heat. The bottom or under side of this plate must be everywhere completely covered and defended from the action of the flame by bricks or tiles. This is very easy to be done; but at the same time, as it requires some care and attention, it is what workmen are very apt to neglect if they are not well looked after. As this plate is very large, if great care be not taken to prevent its being exposed to the flame, it will soon be warped and thrown out of its place. If, instead of casting this plate in one piece, it be formed of two pieces, each 18 inches square, the bad effects produced by the expansion of the metal by heat will be greatly lessened, and this precaution has been taken in most of the register-stoves on these principles that have been put up in London; but by an experiment lately made at Heriot's Hospital, at Edinburgh, I have been convinced that the large plates may be depended on if they are properly set.

I have described the cylinder in which the stewpan or boiler is suspended as being a separate thing. It is right, however, that I should inform the reader that, in almost all cases where register fire-places of this kind have hitherto been put up, this cylinder has been firmly and inseparably united to the stewpan, so much so as to make a part of it, the handle even being attached to this cylinder instead of being joined immediately to the stewpan. The following figure, which represents a vertical section of one of these stewpans and its cylinder, will show how they have hitherto generally been constructed: —

Fig. 51.

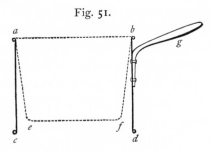

*a*, *b*, *c*, *d*, represents a vertical section of the cylinder, which is 11½ inches in diameter and 8 inches high. Into this cylinder, which is open at both ends, the boiler or stewpan, *a, e, f, d* (which is distinguished by dotted lines), is made to pass with so much difficulty as to require a considerable force to bring it into its place, and not to be in danger of being separated from it by any accidental blow. The handle, *g*, is riveted to the cylinder previously to its being united to its stewpan.

It having been found that this cylinder was liable to become very hot, and even to be destroyed by the heat in a short time if care was not taken to keep the fire low; and it having likewise been found that the heat that made its way upwards, between the outside of the stewpan and the inside of the cylinder, frequently heated the upper part of the stewpan so intensely hot as to cause the victuals cooked in it to be burned to the sides of the stewpan, especially when the stewpan was almost empty, — with a view to remedy both these evils, and at the same time to construct stewpans and saucepans of large dimensions of common sheet tin (tinned iron) which should be more durable, and superior in many respects to those of that material now in common use, some alterations were made in this utensil, which will be easily understood by the help of the following figure : —

Fig. 52.

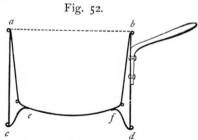

In order to prevent the flame from passing upwards between the saucepan and its cylinder, and occupying the vacant space, *c*, *a*, *e*, this space was enclosed by means of a circular piece of sheet copper, *c*, *e*, *f*, *d*, with a large circular opening in its centre, of the diameter *e*, *f*. This copper, being a little larger in diameter than the cylinder, was firmly attached to it all round by being turned over the same wire, which strengthened and made a finish to the bottom of the cylinder; while the inside edge, *e*, *f*, of this circular perforated sheet of copper, being raised upwards with the hammer about an inch, as it is represented in the figure, the saucepan is made of such a form that, on being brought into its place, its bottom is forced down upon the upper edge of this copper, by which means the empty space between the saucepan and its cylinder is closed up below by the copper, and the flame prevented from entering it. Sheet iron might have been used instead of sheet copper for closing up this space; but copper was preferred to it on account of its not being so liable as iron to be destroyed by the action of the flame.

This contrivance was found to answer so well for preventing the cylinder from being destroyed by heat, that, when it was made of tinned sheet iron (commonly, but improperly, called tin), the tin by which the surface of the iron was covered was not melted by it; and so completely did it prevent the sides of the saucepan from becoming too hot, that a quantity of fluid of any kind, so small as barely to cover the bottom of the vessel, might be boiled in it without the smallest danger of its being burned to its sides.

Having found that the sides of the saucepan were so effectually defended by this contrivance from intense heat, it occurred to me that a saucepan of common tin might perhaps be so constructed as, with this precaution for the preservation of its sides, it might be made to last a great while, which would not only save a considerable expense for kitchen utensils, — tin being much cheaper than copper, — but would also remove the apprehension of being poisoned by any thing injurious to health communicated to the food by the vessel in which it is prepared, which those cannot help feeling who eat victuals cooked in copper utensils, and who know the deleterious qualities of that metal.

Concluding that if I could contrive to prevent the seams or joinings of the tin in a saucepan or boiler from ever coming into contact with the flame of the fire, it could not fail to contribute greatly to the durability of the utensil, I caused the saucepan represented in the foregoing figure to be made of that material. The bottom of this saucepan, *e, f,* was made dishing (instead of being flat, as the bottoms of tin saucepans are commonly made); and, being joined to the body of the saucepan by a strong double seam, the vacuities of the seam, both within and without, were well filled up with solder.

Now as care was taken in adjusting the conical band of copper, *c, e, f, d,* to the bottom of the saucepan, to make its circular opening above, at *e, f,* something less in diameter than the bottom of the saucepan at its extreme breadth, or where it joins the sides or body of the utensil, and also to cause the upper edge of this copper actually to touch the bottom of the saucepan, and even to press against it in every part of its circumference, it is evident that the seam by which the body of the saucepan and its dishing bottom were united was completely covered by the copper, and defended from the immediate action of the fire. It is likewise evident that the side-seams in the body of the saucepan were likewise protected most effectually from all the destructive effects of intense heat; and, if care were taken to cover the outside of the body of the saucepan with a good thick coating of japan to prevent its being injured by rust, there is little doubt but that saucepans so constructed would last a long time indeed.

The cylinder in which the saucepan is suspended might likewise be japanned, both within and without, which would not only preserve it from rust, but would also give it a very neat appearance. All these improvements have been made, and a variety of saucepans constructed on the principles here recommended may be seen in the Repository of the Royal Institution.

*Of the Means that may be employed for using indifferently Saucepans and Boilers of different Sizes, with the same Register-Stove Fire-place.*

Although the diameter below of the cylinder or cone (for it may be either the one or the other) in which the saucepan or boiler is suspended is limited by the diameter of the groove of the circular register-plate in which it stands over the fire, yet the sizes of the cooking utensils used with them may be greatly varied. They may, without the smallest inconvenience, be made either broader or narrower above at their brims than the bottom of the cylinder or cone in which they are suspended; and, with any given breadth above, their depths (and consequently their capacities) may be varied almost at pleasure. When, however, the diameter of one of these boilers, at its brim, is greater than the diameter of the groove of the register-plate of the fire-place, it must be suspended in an inverted hollow cone, and its body must necessarily be made conical.

The following figure shows how a boiler 15 inches in diameter, with a steam-rim (with which the steam-dishes of a 15-inch family boiler may occasionally be used), may be adapted to a register-stove fire-place of the usual dimensions: —

Fig. 53.

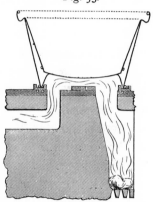

This boiler requires no handle, as its steam-rim may be used instead of a handle in moving it from place to place.

The following figure shows how very small sauce-pans are to be fitted up, in order to their being used with these register-stove fire-places: —

Fig. 54.

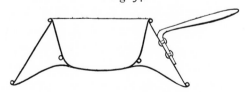

This saucepan is only 6 inches in diameter at its brim, and 3 inches deep. The hollow cone in which it is suspended is about 6 inches in diameter above, $10\frac{1}{2}$ inches in diameter below, and 4 inches in height.

In kitchens of a moderate size it will seldom be convenient to devote more space for stoves for stewpans and saucepans than would be necessary for erecting one register stewing-stove fire-place, which, if the fire-place has only two registers, will heat only two stew-pans or boilers at the same time; but in cooking for a large family it will frequently be necessary to have culinary processes going on at the same time in several stewpans and saucepans. It remains therefore to show how this may be done with the apparatus and utensils just described; and it is certain that this object is so important that any arrangement of culinary apparatus would be essentially deficient and imperfect, which did not afford the means of attaining it completely, and without any kind of difficulty. There are two ways in which it may be done with the utensils above described. A stewpan or saucepan having been placed upon one of the register-plates of the stove till its contents are boiling-hot, it may be removed and placed over a very small fire made with charcoal in a small portable fur-nace resembling a common chafing-dish; or it may be set down upon a circular iron heater, made red-hot, and placed in a bed of dry ashes in a shallow earthen pan. By either of these methods a boiling heat may be *kept up* for a long time in the stewpan; and any common process of boiling or stewing carried on in a very neat and cleanly manner. It must however be remembered that it is only with stewpans and boilers constructed on the principles here recommended, and constantly kept well covered with double covers to prevent the loss of the heat, that the processes of boiling and stewing can be carried on with very small portable furnaces and with heaters; but with these utensils, which are so well calculated to confine the heat, it is almost incredible how small a supply of heat will be sufficient, when the contents of the vessel have previously been made boiling-hot, to keep up that temperature, and carry on any of the common processes of cookery.

In the following figure (Fig. 55) A represents a vertical section of a stewpan, 11 inches wide at its brim and

Fig. 55.

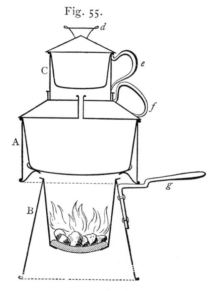

6 inches deep, suspended in its cylinder and placed upon a *portable furnace*, B, which is 7 inches in diameter at its opening above, 11 inches in diameter below, and 9 inches high. A small saucepan, C, for melting butter, is placed on the cover of the stewpan, and is heated by the steam from the stewpan.

This small saucepan is suspended in a cylinder, which serves for confining the steam about it which rises from the stewing-stove.

The cover of this small saucepan is double, and, instead of a handle, it is furnished with a kind of a knob (*d*) formed of a hollow inverted cone of tin, which occasionally serves as a foot for supporting the cover when it is taken off from the saucepan and laid down in an inverted position. This contrivance is designed to prevent the inside of the cover from being exposed to dirt when it is occasionally taken off and laid down. The saucepan is furnished with a handle of the common form (*e*), which is represented in the figure. The handle (*f*) of the stewpan is also shown, and that (*g*) of the portable fire-place.

The following figure is a perspective view of the portable furnace without the stewpan: —

Fig. 56.

In this figure the three horizontal projecting arms are distinctly seen, which serve to support the stewpan. One of these arms, which is longer than the rest, serves as a handle to the furnace.

This little furnace, which is constructed principally of sheet iron, is made double, that part of it which contains the burning charcoal being cylindrical, or nearly so, and being suspended in the axis of a hollow cone, which forms the body of the furnace, and serves as a covering for confining the heat.

The following figure, which represents a vertical section of this furnace through its axis, will give a clear idea of the manner in which it is constructed : —

Fig. 57.

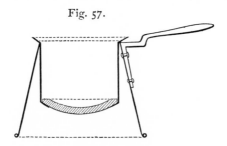

The air is introduced into the fire-place first through a circular hole (represented in the Fig. 56), about $1\frac{1}{2}$ inches in diameter, situated in the side of the hollow cone near its bottom; and from thence it passes up through a small dishing-grate of cast iron which lies at the bottom of the hollow cylinder which contains the burning fuel.   At the upper end of this cylinder there is a narrow rim about half an inch wide, turned outwards, by which the cylinder is suspended in its place; and a similar rim being turned inwards below serves as a support for the dishing-grate.

When this fire-place is used, it will be proper to place it on a flat stone or on a tile; or, what will be still better, to set it in a thin earthen dish.

The same earthen dishes which would be proper for

holding these portable fire-places would also answer perfectly well for holding the cast-iron heaters that may occasionally be used for finishing the processes of cooking that have been begun in stewpans and saucepans heated over the fire of a register-stove, or otherwise made boiling-hot.

The following figure, which represents a vertical section of a stewpan placed over a heater of the kind here recommended, will give a perfect idea of this arrangement : —

Fig. 58.

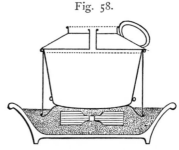

The heater is here represented as lying in a bed of ashes, and there is likewise a thin layer of ashes seen between the top of the heater and the bottom of the stewpan. By the quantity of ashes suffered to remain on the upper surface of the heater, the heat communicated to the stewpan is to be moderated and regulated.

The heater is perforated in its centre by a hole of a peculiar form, which serves for introducing an iron hook, which is used in taking it from the fire and placing it in the earthen dish.

The form of the hook, and the shape of the aperture through which it passes in the heater, may be seen in the following figure.

The circular excavation in the heater, on each side of it, surrounding the hole (which is in the form of the key-hole of a lock) by which the hook is introduced, serves to give room for the hook (or key, as it might be called) to be turned round when the heater is laid

Fig. 59.

upon or against a flat surface. As this excavation, as well as the hole through which the key passes, may be cast with the heater, this arrangement will cause no additional expense.

---

## CHAPTER XI.

*Of the Use of* PORTABLE FURNACES *for culinary Purposes. — Description of a portable Kitchen Furnace, for Boilers, etc., on the common Construction. — Description of a small portable Furnace of cast Iron for heating Tea-kettles, Stewpans, etc. — Description of another of sheet Iron, designed for the same Uses. — Description of a portable Kitchen Furnace of Earthen-ware. — An Account of a very simple Apparatus for cooking used in China.*

IN China and in several other countries, all, or nearly all, the fire-places used in cooking are portable, and real advantages might certainly be derived in many cases from the use of portable kitchen fire-places in this country. Convinced of the utility of this method of cooking, I have taken considerable pains to investigate the subject experimentally, and to ascertain the best forms for the furnaces and utensils necessary in the practice of it.

Portable furnaces for cooking are of two distinct kinds: the one has a fire-place door for introducing the fuel, the other has none; and either of these may or may not be furnished with a tube for carrying off the smoke into the air or into a neighbouring chimney.

When a portable kitchen furnace is constructed without a fire-place door, as often as fuel is to be introduced it will be necessary to remove the boiler, in order to perform that operation. When the boiler is small, that may easily be done; and when the furnace stands out of doors, or on the hearth within the draught of a chimney, or when the fuel used produces little or no smoke, it may be done without any considerable inconvenience. But, if the boiler be large, it cannot be removed without difficulty; and when the furnace is placed within doors, and the fuel used produces smoke or other noxious vapours, the removing of the boiler, though it were but for a moment, would be attended with very disagreeable consequences.

Small portable furnaces without fire-place doors may be used within doors, provided they be heated with charcoal; but it will in that case always be advisable to furnish them with small tubes of sheet iron for carrying off the unwholesome vapour of the charcoal into the chimney. Without such tubes to carry off the smoke, they would not, it is true, be more disagreeable or more detrimental to health than the stoves now generally used for burning charcoal in kitchens; but I should be sorry to recommend an invention to which there appear to me to be so great objections.

I have caused a considerable number of portable kitchen furnaces, of both the kinds above-mentioned, to be constructed; and I shall now give descriptions of such of them as seem to answer best the purposes for which they were designed. They may all be seen at the Repository of the Royal Institution.

A very simple and useful portable kitchen furnace, with its stewpan in its place, is represented by the following figure:—

Fig. 60.

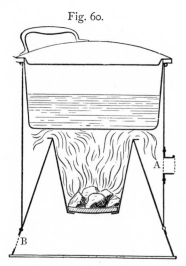

This furnace is made of common sheet iron, and it may be afforded at a very low price. It is composed of a hollow cylinder, and two hollow truncated cones of different sizes. The large cone, which is erect, is closed at its base or lower end. The smaller is inverted, and is open at both ends. This smaller cone is suspended in the larger, by means of a rim about half an

inch wide, which projects outwards from its upper
(larger) end.   A rim of equal width, projecting inwards
at its lower extremity, supports a circular grate, on
which the fuel burns.   The cylinder, which is about
two inches less in diameter than the larger cone at its
base, and which rests upon the surface of that cone,
serves to support the boiler or saucepan.   This cylinder is firmly fixed to the cone on which it rests by
means of rivets, two of which are represented in the
figure.   The upper end of this open cylinder is strengthened, and its circular form preserved, by means of a
strong iron wire, over which the sheet iron is turned.
There is a short horizontal tube (A) on one side of the
cylinder, which is destined for receiving a longer tube
which carries off the smoke.   The air necessary for the
combustion of the fuel is admitted through a circular
hole (B), about $1\frac{1}{4}$ inches in diameter, in the side of
the larger cone near its bottom, and below the joining
of the cone with the cylinder which rests on it.   This
hole for the admission of air should be furnished with
a register, by means of which the fire may be regulated.
The handle of the stewpan is omitted in this plate, as
is also that of the fire-place.   This figure is drawn to
a scale of 8 inches to the inch.

The following figure (which is drawn to a scale of 12
inches to the inch) is a perspective view of one of these
portable furnaces without its stewpan.

A part of the handle of this furnace is seen on the
left hand; and the short tube is seen on the right hand,
that receives another tube (a part of which only is
shown) by which the smoke passes off.

The stewpan represented in the Fig. 60 is supposed
to be made of copper, and to be constructed on the

principles recommended in the seventh chapter of this (tenth) Essay. These portable furnaces are peculiarly adapted to kitchen utensils constructed on those principles, and also to boilers and stewpans with steam-rims, which are not made double; but for **double or**

Fig. 61.

*armed boilers*, stewpans, etc., the furnace must be made in a different manner. The simplest form for portable furnaces adapted to armed boilers is that represented by the Figs. 55, 56, and 57; but I shall now give an account of a furnace of this sort constructed on different and better principles.

The following figure represents a vertical section of a small portable kitchen furnace of *cast iron*.

On examining this figure, it will be found that care has been taken, in contriving this furnace, to divide it in such a manner into parts, and to give to those parts such forms as to render the whole of easy construction. It consists of three principal parts; namely, of the fire-place, A, which is a hollow cylinder, or rather an inverted hollow truncated cone, 7 inches in diameter above measured internally, 4 inches long or high, ending below with a hemispherical hollow bottom, 6 inches in diameter, perforated with many holes for the admission of air.

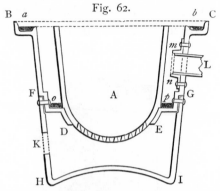

Fig. 62.

This fire-place is suspended in the axis of the furnace by means of the projecting hollow ring, D, E, belonging to the upper and principal piece, B, C, D, E, of the furnace.  At the upper part of this piece there is a circular cavity, *a*, *b*, about 1 inch wide and a quarter of an inch deep, which is destined to receive the lower extremity of the hollow cylinder in which the boiler is suspended.  At L is a circular hole, $1\frac{1}{4}$ inches in diameter, which receives the end of the tube by which the smoke is carried to the chimney.  A part of this tube, which is of sheet iron, is represented in the figure.  To give it a more firm support in its place, there is a short tube, *m*, *n*, of cast iron, which projects inwards into the furnace about $\frac{5}{8}$ of an inch.  This short tube is cast with a flange, and it is fastened to the inside of the piece which constitutes the upper part of the body of the furnace by means of three or four rivets.  Two of these rivets are distinctly represented in the figure.

The lower part of the body of the furnace consists of the piece, F, G, H, I, and it is fastened to the upper part by means of rivets, two of which are seen at F and at G. In one side of this lower part there is a circular hole at K, about $1\frac{1}{4}$ inches in diameter, which serves for the admission of air, and which is furnished with a register-stopper. The bottom of this furnace, instead of being made flat, is spherical, projecting upwards; which form was chosen in order to prevent as much as possible the heat from the fire from being communicated downward. This furnace will require no handle, as its projecting brim will serve instead of one.

It will be observed that all the pieces of which this furnace is composed are of such forms that the moulds for casting them will readily deliver from the sand; and that circumstance will contribute greatly to the lowness of the price at which this most useful article of kitchen furniture may be afforded.

The perforated cast iron bowl, A, which constitutes the fire-place, is not confined in its place, and its form and its position are such that its expansion with heat can do no injury to the outside of the furnace.

When the two pieces which form the body of the furnace are fastened together, their joinings may be made tight with cement.

A little fine sand should be put into the hollow rim, $a$, $b$, of the furnace, in order that it may be perfectly closed above by the lower end of the hollow cylinder of its boiler; and a little sand or ashes may be thrown upon the bottom of the circular cavity, $o$, $p$, into which the smoke descends before it goes off by the tube, L, into the chimney. This last precaution will prevent the air from making its way upwards from the ash-pit directly into the cavity, $o$, $p$, occupied by the smoke, without passing through the fire-place.

The register-stopper to the opening, K, into the ash-pit, may be constructed on the same principle as that of the blowpipe of a roaster. One of these stoppers is represented on a large scale in the Fig. 17, at the end of the second part of this (tenth) Essay; or, what will be still more simple and quite as good, the admission of the air may be regulated by a register like that represented in the preceding Fig. No. 61.

This portable kitchen furnace will answer a variety of useful purposes; and, if I am not much mistaken, it will come into very general use. It is cheap and durable, and not liable to be broken by accidents or put out of order; and it is equally well adapted for every kind of fuel. No particular care or attention is required in the management of it, and it is well calculated for confining heat, and directing it.

As the fire-place belonging to this furnace is nearly insulated, and as it contains but a small quantity of matter to be heated, a fire is easily and expeditiously kindled in it; and the fuel burns in it under the most favourable circumstance.

It will be found extremely useful for boiling a tea-kettle, especially in summer, when a fire in the grate is not wanted for other purposes; and, when the tea-kettle is constructed on the principles that will presently be described, a very small quantity indeed of fuel will suffice.

But the most important use to which these portable furnaces can be applied is most undoubtedly for cooking for poor families. I have hinted at the probable utility of a contrivance of this kind in some of my former publications; but since that time I have had opportunities of examining the subject more attentively, and of ascertaining the fact by the test of actual experiment.

As the subject strikes me as being of no small degree of importance, I shall make no apology for enlarging on it, and giving the *most particular account* of several kinds of *portable kitchen furnaces.*

That just described (of cast iron) is, it is true, as perfect in all respects as I have been able to make it, and will probably be found to be quite as economical and as useful as any that I shall describe ; but cast iron is not everywhere to be found, and, even where foundries are established for casting it, moulds must be provided, and these are expensive, and not easy to be had. As it is probable that some persons may be desirous of being provided with portable furnaces of this kind, who may not have it in their power to procure them of cast iron, I shall now show how they may be constructed (by any common workman) of sheet iron, and also how they may be made of earthen-ware.

*Of small portable Kitchen Furnaces constructed of sheet Iron.*

The following figure represents a vertical section of one of these furnaces, drawn to a scale of 6 inches to the inch.

The construction of this furnace will be easily understood from this figure. The circular hollow horizontal rim, *a, b*, which I shall call the *sand-rim*, is $8\frac{8}{10}$ inches in diameter within, and $12\frac{4}{10}$ inches in diameter without. Its width at its bottom, which is flat, is just 1 inch. Its sides are sloping and of different heights : that which is towards the centre of the furnace is $\frac{1}{4}$ of an inch high, but the side which is outwards is $\frac{1}{2}$ an inch in height.

The sand-rim is confined and supported in its place by being fastened, by means of rivets or otherwise, to an inverted hollow truncated cone, *c, d, e, f,* which forms the upper part of the body of the furnace. This inverted cone, which is turned over a strong circular iron wire at its upper edge, *c, d,* is $12\frac{4}{10}$ inches in diam-

Fig. 63.

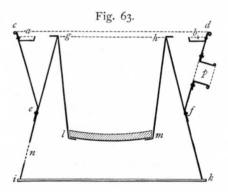

eter above measured within the wire, and $5\frac{4}{10}$ inches in height measured from *c* to *e* or from *d* to *f,* and is $9\frac{4}{10}$ inches in diameter from *e* to *f,* where it is fastened to the erect hollow truncated cone, *g, h, i, k.*

This last-mentioned erect cone, which is closed below by a circular plate of sheet iron, forms the lower part of the body of the furnace. It is 7 inches in diameter above, 12 inches in diameter below, and its perpendicular height is just 9 inches. Its sloping side, *g, i,* measures about $9\frac{6}{10}$ inches.

The *fire-place* of this little portable furnace is an inverted hollow truncated cone, *g, h, l, m,* which is 7 inches in diameter above, at *g, h,* and $5\frac{1}{2}$ inches in diameter below, at *l, m;* and its length is $6\frac{1}{2}$ inches, measured from *g* to *m*. This conical fire-place has a flat rim above, which is $\frac{1}{2}$ an inch wide, and turned outwards; and another below of equal width which is turned inwards. The first serves to suspend it in its place, the second serves to support its circular grate on which the fuel burns.

The air is admitted into the fire-place through a hole, *n*, about $1\frac{1}{4}$ inches in diameter, in the side of the furnace. This aperture must be furnished with a register similar to that shown in the Fig. 61.

The provision for carrying off the smoke is similar in all respects to that used in the portable furnace above described, constructed of cast iron; and it will easily be understood, from a bare inspection of the Fig. 63, without any farther explanation.

Having shown how this portable kitchen furnace may be constructed of cast iron, and also how it may be made of sheet iron, I shall now show how it may be made partly of cast iron and partly of sheet iron. A fire-place of cast iron, like that represented in the Fig. 62, may be used in a furnace of sheet iron; but, when this is done, the fire-place must be cast with a projecting rim above, in order that it may be suspended in its place. The sand-rim may likewise be of cast iron, and it may be fastened to the inverted hollow cone, *c*, *d*, *e*, *f*, by rivets.

The short tube, *p*, which serves to support the tube which carries off the smoke, may also be made of cast iron, and it may be fastened to the outside of the furnace by three rivets. As it may be made of such a form that its mould will deliver from the sand, it will cost less when made of cast iron than when made of sheet iron; and it will have another advantage, — its form on the inside will be more regular, and it will be better adapted on that account for receiving the end of the tube, which it is designed to receive. Its length need not exceed 1 inch or $1\frac{1}{2}$ inches, and its internal diameter may be about $1\frac{1}{2}$ inches at its projecting extremity, and something less at its other end, where it joins the side of the furnace.

*Of small portable Kitchen Furnaces constructed of Earthen-ware.*

The following figure represents a furnace of this kind (of earthen-ware) destined for heating boilers of the same kind and of the same dimension as those proper to be used with the two (iron) furnaces last described : —

Fig. 64.

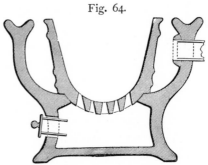

This figure represents a vertical section of the furnace, drawn to a scale of 6 inches to the inch; and it gives an idea so clear and satisfactory of the form of this furnace that a detailed description of it would be superfluous.

The fire-place is distinct from the body of the furnace, and its form and position are such that it cannot crack and injure the body of the furnace by its expansion with heat. It resembles very much the cast iron

fire-place just described, and the same principles regulated the contrivance of both of them.   It should be
bound round with iron wire, in order to hold it together,
in case it should crack with the heat of the fire.   Two
places for the wire, one near its brim and the other
lower down, are shown in the figure.

The aperture by which the air enters the ash-pit is
closed by a register-stopper, represented in the figure,
or a conical stopper of earthen-ware may be used for
that purpose.

If such earths are used in constructing these small
portable furnaces as are known to stand fire well, there
is no doubt but these furnaces may, with proper usage,
be made to last a great while; and, for confining heat,
they are certainly preferable to all others.

The portable kitchen furnaces in China are all constructed of earthen-ware; and no people ever carried
those inventions which are most generally useful in
common life to higher perfection than the Chinese.
They, and they only, of all the nations of whom we
have any authentic accounts, seem to have had a just
idea of the infinite importance of those improvements
which are calculated to promote the comforts of the
lowest classes of society.

What immortal glory might any European nation
obtain by following this wise example!

The emperor of China, the greatest monarch in the
world, who rules over one full *third part* of the inhabitants of this globe, condescends *to hold the plough*
himself one day in every year.   This he does, no doubt,
to show to those whose example never can fail to influence the great bulk of mankind how important that art
is by means of which food is provided.

Let those reflect seriously on this illustrious example of provident and benevolent attention to the wants of mankind who are disposed to consider the domestic arrangements of the labouring classes as a subject too low and vulgar for their notice.

If attention to the art by which food is provided be not beneath the dignity of a great monarch, that art by which food is prepared for use, and by which it may be greatly *economized*, cannot possibly be unworthy of the attention of those who take pleasure in promoting the happiness of mankind.

As the implements used in China for cooking are uncommonly simple, it may perhaps be amusing to the reader to be made acquainted with them.   They consist of the two articles represented below : —

Fig. 65.

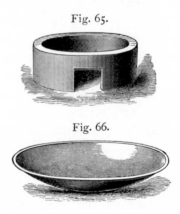

Fig. 66.

This Fig. 65, which is made of earthen-ware, is the fire-place, which is set down on the ground.   The shallow pan, represented by the Fig. 66, is of cast iron, and serves for every process of Chinese cookery.   It is cast very thin, and, if by any accident a hole is made in it, their itinerant tinkers mend it by filling up the hole, which they do with so much dexterity that scarcely a mark is left behind.

When the dinner consists of several dishes, they are all cooked in this pan, one after the other; and those which are done first are kept warm till they are sent to table.

I leave it to the ingenuity of Europeans to appreciate these specimens of Chinese industry.

But to return from this digression to our portable kitchen furnaces. Although these furnaces are peculiarly adapted for heating boilers and stewpans that are *armed*, yet boilers on the common construction, or such as are not suspended in cylinders, may easily be used with them. When this is to be done, a detached hollow cylinder or cone must be used in the manner described in the preceding chapter, and represented in the Fig. 50. This cylinder or cone (which may be constructed either of sheet iron, of cast iron, or of earthen-ware) must be about an inch higher than the boiler is deep, with which it is to be used ; and just so wide above as to admit the boiler to be suspended in it by its circular rim. Its diameter below must be such as to fit the sand-rim, in which it must stand when it is used.

----

# CHAPTER XII.

*Of the Construction of* TEA-KETTLES *proper to be used with Register-Stoves and portable Kitchen Furnaces. — These Utensils may be constructed of Tin, and ornamented by Japanning and Gilding. — When*

*they are properly constructed and managed, they may
be heated over a small portable Furnace in a very
short Time, and with a surprisingly small Quantity
of Fuel.—Descriptions of four of these Tea-kettles of
different Forms and Sizes. — Description of several
very* SIMPLE *and* CHEAP STEWPANS *for portable Furnaces.—Description of a* STEWPAN *of* EARTHEN-WARE
*on an improved Construction. — This will probably
turn out to be a most useful Utensil for cooking with
portable Furnaces.*

AS tea-kettles are so much used in this country,
and as they occasion so great a consumption of
fuel (a large fire being frequently made in a grate or
kitchen range, morning and evening, for the sole purpose of heating a few pints of water to make tea), the
saving of this unnecessary trouble and expense is an
object deserving of attention. And in doing this it
will be possible to improve very essentially the forms of
tea-kettles in several respects, and at the same time to
render their external appearance more neat and cleanly.
If the forms I shall recommend should not happen
to please at first sight, it should be remembered that
utility, cleanliness, and wholesomeness are objects of
more importance in cases like that in question than
mere elegance of form; and, after all, I am not sure
whether the forms I shall propose are not in reality
quite as elegant as those with which they will be compared. They will, no doubt, at first sight appear
uncouth to many persons, but the eye will soon become
accustomed to them; and their superior cheapness,
cleanliness, and usefulness will in the end procure
them that preference which they deserve. They may,

no doubt, be constructed of the most elegant forms, on the principles I shall recommend; but I shall confine my descriptions to such forms as are most simple, and of the easiest and least expensive construction, leaving it to those to beautify the article whose business and interest it is to set off their goods to the best advantage.

The following figure represents a tea-kettle of the simplest form, suited to a register kitchen stove, or to a portable furnace such as has just been described : —

Fig. 67.

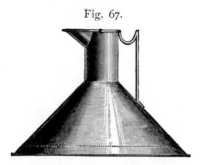

This tea-kettle is constructed of tin, and it may be japanned on the outside to prevent its rusting, and to give it an elegant and cleanly appearance. Its bottom, which is 11 inches in diameter, is not flat, but it is raised up about half an inch in the manner pointed out by a dotted line. The body of this tea-kettle is of a conical form, ending above in a cylinder, 3 inches in length and 2 inches in diameter. The spout, which resembles that of a coffee-pot, is situated at the top of this cylinder; and it has a flat cover, fastened by a hinge, which prevents dust or soot from falling into it when it stands on the hearth. When this tea-kettle is put over the fire, it should not be filled higher than to the top of the cone, or lower end of the cylinder, otherwise it will be liable to boil over. The kettle so filled will contain 4 pints of water; and, if it be heated over one of the small portable furnaces described in the foregoing chapter, it may be made to boil in about 10 minutes, with $6\frac{1}{2}$ oz. of dry wood, which, at the price at which wood is commonly sold in London, would cost $\frac{3}{8}$ of a farthing.*

The tea-kettle represented by the following figure is rather more complicated, but still its form is more simple, and more advantageous in several respects than those which are in common use, and it is well adapted for the fire-places we have recommended. It is drawn to a scale of 6 inches to the inch.

Fig. 68.

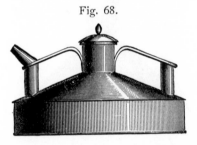

This kettle has two handles, each of which is supported on the outside, or near the circumference of the kettle, by a small vertical tube, $\frac{3}{4}$ of an inch in diameter and $1\frac{3}{4}$ inches in height. That on the left hand is open, and forms a part of the spout; but that on the right hand is closed at both ends. The bottom of this kettle, also the bottoms of those represented in the two following figures, like that of the last (Fig. 67), is not flat, but is raised up about half an inch above the level of the lower part of the cylindrical sides of the kettle.

* One pint of water only being put into this tea-kettle, over a very small wood fire, made in the portable furnace represented in the foregoing Fig. 63 (see page 310), it was heated and made to boil *in two minutes and a half.*

This kettle holds about 3 quarts of water, which can be made to boil with the combustion of $9\frac{1}{2}$ oz. of wood.

The following kettle holds about 1 gallon, and may be made to boil with $\frac{3}{4}$ lb. of wood, which would cost just $\frac{3}{4}$ of a farthing : —

Fig. 69.

The following kettle is not essentially different from those two last described, except in the form of its handle. It holds about 3 quarts.

Fig. 70.

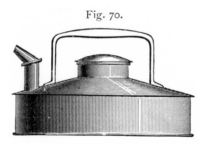

The cylindrical opening of this kettle above, where the water is introduced, is considerably wider than those in the two foregoing figures. It was made wider because it was necessary to make it lower, in order to make room for the hand without raising the handle too high. When this part of a tea-kettle is made very narrow, it must be made high to afford room for the expansion of the water with heat, and prevent the kettle from boiling over. These kettles should never be filled higher than to the level of the lower part of this cylindrical space, otherwise there will be danger of their boiling over.*

It will be observed that the cover of this tea-kettle projects a little beyond the cylindrical opening to which it belongs. This projection serves instead of a handle in removing and replacing the cover. The cover of a tea-kettle is usually furnished with a knob for that purpose; but these knobs are in the way when the kettle is lifted up by its handle, unless the handle be made much higher than otherwise would be sufficient.

It has, no doubt, already been remarked by the reader that all the tea-kettles here recommended are of forms that are perfectly easy to be executed in tin. There are several reasons which have induced me to give a decided preference to that material for constructing culinary utensils. It is not only wholesome, — which copper is not, — but it is also very cheap, and easy to be procured in all places, and it is easily worked. It is moreover light and strong, and not liable to be injured by accidents; and if measures be taken to prevent the effects of rust it is very durable.

The four tea-kettles represented in the four last figures are all particularly designed to be used with the portable furnaces described in the last chapter; and for that purpose they are well calculated, although they are not suspended in cylinders. They may likewise be used with the register kitchen stoves described

* I find, by experiments made since the above was written, that tea-kettles of this kind should never be filled above two thirds full, otherwise they will be very apt to boil over.

in the tenth chapter of this Essay. As their bottoms are raised up, and as their diameters are such that their conical or vertical sides enter into and fit the sand-rims of those furnaces and stoves, the heat is effectually confined under them; and their outsides, not being exposed either to flame or to smoke, may be japanned, and they may easily be kept so clean as to be fit to be placed upon a table, over a lamp, or upon a heater placed in a shallow dish of china or earthen-ware. They are even capable of being elegantly ornamented by gilding or painting, or both.

They are likewise well calculated for being heated by a lamp; and if an Argand's lamp be used for that purpose they may be made to boil in a short time and at a small expense. Placed on a handsome tripod on a table, with an elegant Argand's lamp under it, one of these kettles, handsomely ornamented by japanning and gilding, would make no mean appearance, and would cost much less than the commonest tea-urn that could be bought.

But it is not solely for making tea that these kettles will be found useful: they will answer perfectly well for boiling water for many other purposes; and, if portable kitchen furnaces should come into use, boiling-hot water will often be wanted for filling saucepans and stewpans; and no utensil can be better contrived for heating and boiling water over a portable kitchen furnace than these kettles.

In constructing them, care should be taken to fill all their seams well with solder, which, by covering the naked edges of the iron, will contribute more than any thing to the prevention of rust and the durability of the article; and they should likewise be well japanned on the outside in every part except the bottom, which should not be japanned.

The reason why I have not made these tin tea-kettles double is this: Tea-kettles are commonly used merely for *making water boil*, which, with the kettles here recommended, can be done *in a very short time*, consequently much heat cannot possibly be lost during that process in consequence of the top and sides of the kettle being exposed naked to the cold air of the atmosphere. Were these utensils designed for *keeping water boiling-hot* a great length of time, the case would be very different; and then it might be well worth while to make them double, in order more effectually to confine the heat in them.

The *saving of time* in making them boil by making them double would be very trifling indeed, for till the water has become very hot there is but little loss of heat through the sides and top of the kettle; the communication of heat being rapid in proportion as the temperature of the hot body is high compared with that of the colder body into which the heat passes.

If a tea-kettle filled with water at the temperature of the atmosphere at the time, on being put over a fire, be brought to boil in 10 minutes, it will, during that time, have lost only half as much heat as it will lose in the next 10 minutes, if it be kept boiling-hot during that time.

All these kettles are of such forms as will render it very easy to cover them, should it be thought advisable to make them double; and by covering them with plated or gilt copper they may be made very elegant at a small expense.

*Of the Construction of cheap Boilers and Stewpans to be used with small portable Kitchen Furnaces.*

The best boilers and stewpans that can be used with these furnaces are undoubtedly those which were described in the tenth chapter of this Essay; but utensils on a simpler construction may be made to answer very well, and may perhaps be preferred by many on account of their cheapness.

The following figure represents a vertical section of a stewpan on a much more simple construction than any of those already described: —

Fig. 71.

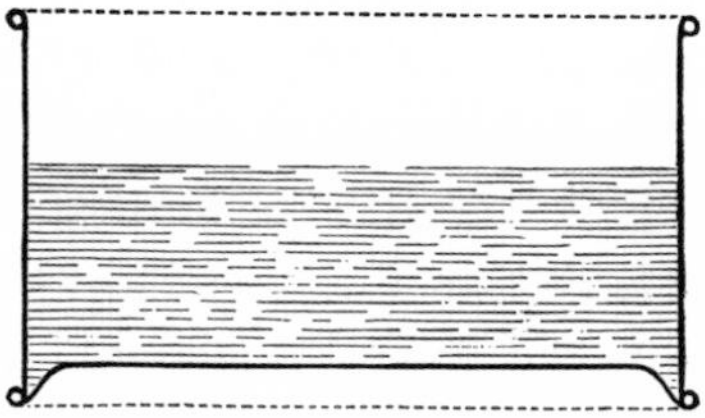

This stewpan (which is drawn to a scale of 6 inches to the inch) being of a proper diameter below to fit the sand-rim of the portable furnace, and its bottom being raised up about half an inch in order to allow its vertical sides to descend into that sand-rim, it is plain that it may be used with the furnace in the same manner as the tea-kettles just described are used with it. It may likewise be used with the register-stoves described in the tenth chapter of this Essay.

In order that this stewpan may the more easily be kept clean, the joinings of its bottom and sides should be well filled up on the inside with solder.

The following figure represents another and smaller

stewpan, constructed on the same principles with that just described and designed for the same use : —

Fig. 72.

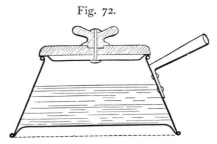

The diameter of this stewpan below is the same as that of the last.   This is necessary, in order that it may fit the sand-rim of the same register-stove or portable furnace; but its diameter above is much less, and it is also less deep, consequently its capacity is much smaller. The cover of this stewpan is of wood lined with tin.   It is in all respects like that represented by the Fig. 35 (see Chapter VII. of this Essay, page 254).   Both these stewpans are supposed to be constructed of tin, but they might be made of tinned copper.   The handle of the stewpan represented by the Fig. 71 is omitted.

The following figure represents a vertical section of a double or armed stewpan on a very simple construction : —

Fig. 73.

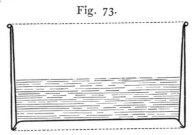

The stewpan (which is drawn to a scale of 6 inches to the inch) is supposed to be made of tin, and it is supposed to be turned over a wire at its brim. The cylinder by which it is surrounded is of sheet iron, and the stew-pan and the cylinder are fastened together by the former being driven into the latter with some degree of force, and sticking in it above where they come into close contact. The lower edge of the cylinder being turned inwards forms a narrow rim on which the lower end of the stewpan rests.

*Of the Construction of Stewpans of* EARTHEN-WARE *and* PORCELAIN, *to be used with Register-Stoves and portable Kitchen Furnaces.*

The following figure shows how, by means of a hoop or cylinder of sheet iron, a stewpan or saucepan of earthen-ware or of porcelain of a suitable form and size may be fitted to be used with a register kitchen stove or portable furnace : —

Fig. 74.

This figure is drawn to a scale of 9 inches to the inch. The form of the lower part of the stewpan is pointed out by a dotted line. The top and the bottom of the cylinder of sheet iron are both turned over circular iron wires. The handle of this stewpan is of iron, and it is fixed to the cylinder by rivets. The stewpan is firmly fastened to its metallic hoop or cylinder, first, by making this cylinder of a proper size to fit it; and, secondly, by wedging it both above and below with very thin wedges made of narrow pieces of sheet iron, and by filling up the vacuities above and below with good cement.

The cover of this stewpan, which is of earthen-ware (or porcelain), is made of a peculiar form. It has a kind of foot instead of a handle, which serves for supporting it when it is taken off from the stewpan and laid down in an inverted position. By means of this simple contrivance it is rendered less liable to be dirtied on the inside and of communicating dirt to the victuals.

If an earthen stewpan of the form represented in this figure be made of good materials, — that is to say, of a proper mixture of the different earths well worked, — and if its bottom be made thin and of equal thickness in every part of it that is exposed to the fire, there is little doubt, I think, of its standing the heat of a register-stove or of a small portable kitchen furnace; and, if this should be the case, I should certainly never think of recommending any other kitchen utensils in preference to these.

It appears to me to be very probable that unglazed Wedgewood's ware would be as good a material as could be found for these stewpans. The intelligent gentleman who directs Mr. Wedgewood's manufactory caused several of them to be made after drawings which I gave him, and those I found, upon trial, to answer very well.

If it should be found that kitchen utensils, constructed and fitted up, or mounted, on the principles here pointed out, should answer as well as there is reason to expect, as nothing would be easier than to make earthen boilers with *steam-rims* and to form *steam-dishes* of earthen-ware to fit them, every utensil for cooking, by *boiling* and *stewing*, might be constructed of that most cleanly, most elegant, and most wholesome material, — *earthen-ware.*

I hesitated a long time before I resolved to publish this last observation; for, however anxious I am to promote useful improvements, and especially such as tend to the preservation of health and the increase of rational enjoyments, it always gives me pain when I recollect how impossible it is to introduce any thing new, however useful it may be to society at large, without occasioning a temporary loss or inconvenience to some certain individuals, whose interest it is to preserve the state of things *actually existing.*

It certainly requires some courage, and perhaps no small share of enthusiasm, to stand forth the voluntary champion of the public good; but this is a melancholy reflection, on which I never suffer my mind to dwell. There is no saying what the consequences might be, were we always to sit down before we engage in a laudable undertaking and meditate profoundly upon all the dangers and difficulties that are inseparably connected with it. The most ardent zeal might perhaps be damped and the warmest benevolence discouraged.

But the enterprising seldom regard dangers, and are never dismayed by them; and they consider difficulties but to see how they are to be overcome. To them *activity* alone is life, and their glorious reward the consciousness of having done well. Their sleep is sweet when the labours of the day are over; and they await with placid composure that rest which is to put a final end to all their labours and to all their sufferings.

## CHAPTER XIII.

*Of cheap Kitchen Utensils for the Use of the Poor. —*
*The Condition of the lower Classes of Society cannot*
*be improved without the friendly Assistance of the*
*Rich. — They must be* TAUGHT *Economy, and they can-*
*not be instructed by Books, for they have not Leisure*
*to read. — Advice intended for their Good must be*
*addressed to their benevolent and more wealthy Neigh-*
*bours. — An Account of the Kitchen Utensils of the*
*poor itinerant Families that trade between Bavaria*
*and the Tyrol. — These Utensils were adopted by the*
*Bavarian Soldiers. — An Account of some Attempts*
*that were made to improve them. — Description of*
*a very simple closed Fire-place constructed with seven*
*loose Bricks. — How this Fire-place may be improved*
*by using three Bricks more, and a few Pebbles. —*
*Description of a very useful* PORTABLE KITCHEN
BOILER *of cast Iron, suitable for a small Family.*
*— An Account of a very simple Method of* COOKING
WITH STEAM, *on the Cover of this Boiler. — Descrip-*
*tion of a* STEAM-DISH *of Earthen-ware or of cast*
*Iron, to be used with this Boiler. — Description of*
*a Boiler still more simple in its Construction, proper*
*to be used with a small portable Kitchen Furnace. —*
*The cooking Apparatus here recommended for the*
*Use of the Poor may, with a small Addition, be*
*rendered serviceable for warming their Dwellings*
*in cold Weather.*

AMONGST the great variety of enjoyments which
riches put within the reach of persons of fortune

and education, there is none more delightful than that which results from doing good to those from whom no return can be expected; or none but gratitude, respect, and attachment. What exquisite pleasure then must it afford to collect the scattered rays of useful science, and direct them *united* to objects of general utility! to throw them in a broad beam on the cold and dreary habitations of the poor, spreading cheerfulness and comfort all around!

Is it not possible to draw off the attention of the rich from trifling and unprofitable amusements, and engage them in pursuits in which their own happiness and reputation and the public prosperity are so intimately connected? What a wonderful change in the state of society might, in a short time, be affected by their *united efforts!*

It is hardly possible for the condition of the lower classes of society to be essentially improved without that kind and friendly assistance which none can afford them but the rich and the benevolent. They must be *taught*, and who is there in whom they have confidence that will take the trouble to instruct them? They cannot learn from books, for they have not time to read; and, if they had, how few of them would be able from a written description to comprehend what they ought to know! If I write for their instruction, it is to the rich that I must address myself; and, if I am not able to engage *them* to assist me, all my labours will be in vain. But to proceed.

In contriving kitchen utensils for cottagers, two objects must frequently be had in view, — viz., the cooking of victuals and the warming of the habitation; and as these objects require very different mechanical arrangements, some address will be necessary in combining them.

Another point to which the utmost attention must be paid is to avoid all complicated and expensive machinery. Instruments for general use should be as simple as possible; and such as are destined for the use of those who must earn their daily bread by their labour should be cheap, durable, and not liable to accidents, or to be often in want of repairs.

As food is more indispensably necessary than a warm room, and as the most common process of cookery is boiling, I shall first show how that process may be performed in the most economical manner possible, and shall then point out the means that may be used for rendering the kitchen fire useful in warming the room in which cookery is carried on.

One of the cheapest utensils for cooking for a family that ever was contrived is, I verily believe, that used by the itinerant poor families that trade between Bavaria and the Tyrol, bringing raisins, lemons, etc., from the south side of the mountains (which they transport in light carts drawn by themselves) and carrying back earthen-ware.

As these poor people have no fixed abode, and never stop at an inn or other public-house, but, like the gypsies in this country, sleep in empty barns and under the hedges by the road-side, they carry with them in their cart all that they possess; and among the rest the whole of their kitchen furniture, which consists of *one single article,* — a deep frying pan of hammered iron, with a short iron handle.

In this they bake their cakes, boil their brown soup, make their hasty pudding, stew their greens, fry their meat, and in short perform every process of their cookery; and, when their victuals are done, their boiler serves them for a dish, which, being placed on the ground, the family sit round it, each individual capable of feeding himself being provided with a wooden spoon.

This is precisely the same kind of kitchen utensil as that used by the Bavarian wood-cutters when they go into the mountains to fell wood; and it is likewise used by many poor families in the Tyrol and in Bavaria.

These broad stewpans, with the addition of a tripod of hammered iron, were adopted many years ago in Bavaria, for the use of the soldiers in barracks; and they still continue to be used by them. Some successful attempts to improve them have, however, lately been made, and it was the experiments which led to those improvements that first induced me to turn my attention to this useful article of kitchen furniture.

Before I proceed any farther in my account of these shallow pans, and of the improvements of which they have been found to be capable, it may perhaps be proper to give an account of the manner in which they are constructed, and of the price at which they are sold.

All those which are used in Bavaria come from the Tyrol or from Styria, where there are considerable manufactories of them; and they are sold at Munich by wholesale at 22 kreutzers (about 7½*d.* sterling) the pound, Bavarian weight, which is at the rate of 6*d.* sterling per lb. avoirdupois weight.

One of these pans of large dimensions, — namely, 18 inches in diameter above or at its brim, 15 inches in diameter below, and 4 inches deep, — bought at an iron-

monger's shop at Munich, cost me three shillings
sterling.

In manufacturing these pans, five of them, one placed
within the other, are brought under the hammer at the
same time; and, in being hammered out and brought
to their proper form and thickness, they are frequently
heated red-hot. When they come from the hammer,
they are carried to the lathe and are turned on the
inside, and made clean and bright, and their edges are
turned and made even. They are then packed up one
within the other, or in nests (as these parcels are called),
and are sold by weight.

The following figure represents one of these pans in
its most simple state, placed on three stones, over a
fire made with small sticks of wood on the ground in
the open air: —

Fig. 75.

The pan used by the Bavarian soldiers — which, as I
just observed, is placed on a tripod or trivet of iron —
is about 20 inches in diameter above, 16 inches in
diameter below, and $4\frac{1}{2}$ inches deep.

As a great part of the heat generated in the combustion of the fuel that is burned under this pan escaped
by its sides, to prevent in some measure this loss, I
enclosed the pan in a circular hoop or cylinder of sheet
iron. The diameter of this hoop was just equal to the
diameter of the pan above or at its brim, and its height

or width was 6 inches, and the upper part of it was fastened by rivets to the upper part or brim of the pan. This alteration, and a double cover fitted to the pan which prevented the heat from being carried off by the cold air of the atmosphere from the broad sur-face of the hot liquid in the pan, produced a saving of considerably more than half the fuel, even when this fuel — which was dry pine wood — was burned on the hearth or on the ground in the open air, and no means were used for confining the heat on either side. But the saving was still greater when the fire was made in a closed fire-place.

For a pan of this kind of 14 or 15 inches in diameter at its brim, a very good temporary fire-place may be con-structed in a moment, and almost without either trouble or expense, merely with seven common bricks. Six of them, laid down upon the hearth in pairs one upon the other in the manner represented in the following figure,

Fig. 76.

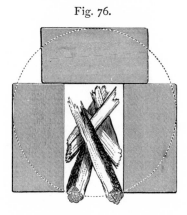

form the fire-place; and the seventh, placed edgewise, serves as a sliding door to close this fire-place in front more or less, as shall be found best.

This little fire-place, which is better calculated for wood or for turf than for coals, is represented filled with fire-wood ready to be kindled, and a dotted circular line shows where the bottom of the circular hoop of sheet iron (in which the pan is suspended) should be set down upon the top of the three bricks which are uppermost.

If, in constructing this fire-place, its walls be made higher by using nine bricks instead of six (laid down flat upon one another by threes), and if a few loose pebbles or stones of any kind, about as large as hens' eggs, be put into it under the fuel, these additions will improve it considerably. The fuel being laid upon these pebbles instead of lying on the hearth or on the ground, the air necessary for its combustion will the more readily get under it, which will cause the fire to burn brighter and more heat to be generated.

These small stones will likewise serve other useful purposes. They will grow very hot, and when they are so they will increase the violence of the combustion and the intensity of the heat; and, even after the fuel is all consumed, they will still be of use by giving off gradually to the pan the heat which they will have imbibed.

Savages, who have few implements of cookery, make great use of heated stones in preparing their food; and civilized nations would do wisely to avail themselves oftener than they do of *their* ingenious contrivances.

I have already mentioned that a considerable saving of fuel was made in consequence of furnishing the broad and shallow boilers of the Bavarian soldiers with double covers; but for boilers of this kind, that are destined for poor families, I would recommend wooden or earthen dishes, turned upside down, instead of these double covers; which dishes may also be used for serving up the victuals after it is cooked. By this contrivance an article necessary in housekeeping will be made to serve two purposes; and, besides this advantage, as a deep bowl or platter turned upside down over the shallow boiler will leave a considerable space above the level of the boiler, which, as steam is lighter than air, will always be filled with hot steam when the water in the shallow pan is boiling, notwithstanding that the joinings of this inverted dish with the rim of the pan will not be steam-tight, a piece of meat much larger than could be covered by the water in this shallow pan might be cooked in it, or potatoes or greens, placed above the surface of the water in the pan, might be cooked in steam.

The following figure, which represents a vertical section of one of these shallow iron boilers, 14 inches in diameter above, surrounded by a cylindrical hoop of sheet iron for confining the heat, and covered by an inverted earthen dish, will give a clear idea of the proposed arrangement: —

Fig. 77.

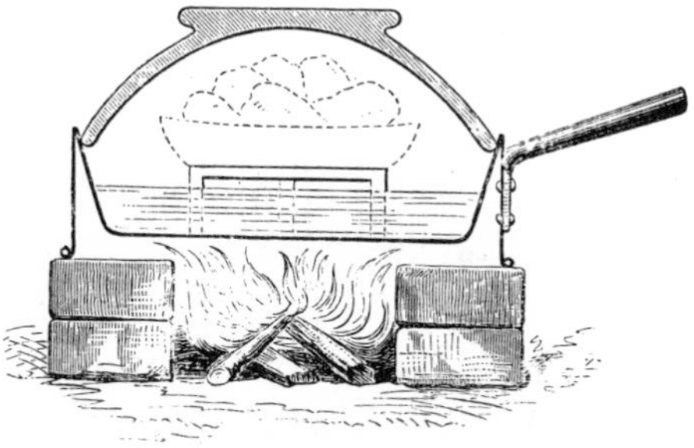

The fire-place represented in this figure is that shown in the preceding figure (Fig. 76), and is constructed of six loose bricks. The brick which occasionally serves to close the opening into the fire-place in front is not shown.

A shallow dish is represented (by dotted lines) standing on a small tripod above the surface of the water in the boiler and filled with potatoes, which are supposed to be boiled in steam.

The earthen dish which covers the boiler is represented with a small projection like the foot which is frequently given to earthen dishes. This projection serves instead of a handle when the dish is placed upon, or removed from, the boiler.

This I believe to be the cheapest contrivance that can be used for cooking victuals for a poor family, especially when the durability of the utensil is taken into the account, and also the small quantity of fuel that is required to heat it. The following contrivance will, however, be found more convenient and not much more expensive.

*Description of a very useful portable Kitchen Boiler of cast Iron, suitable for a small Family.*

The form of this boiler is such that it may easily be cast, and consequently it may be afforded at a low price; and it is equally well calculated to be used with one of the small temporary fire-places just described, constructed with six or with nine loose bricks, or to be heated over one of the small portable kitchen furnaces, of which an account has been given in Chapter XI. It may be made of any dimensions, but the size I would recommend for a small poor family is that indicated by the following figure, which is drawn to a scale of 6 inches to the inch.

This boiler is 10½ inches in diameter above on the inside of the steam-rim, 9½ inches in diameter below, and 8½ deep, measured from the top of the inside of

Fig. 78.

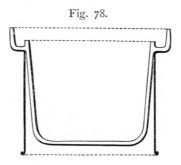

the steam-rim; consequently it will hold about 3 gallons.   Its greatest diameter at its brim is 13½ inches, and total height to the top of its steam-rim is 9¾ inches.

The hollow cylinder of sheet iron in which this boiler is suspended, and which confines the heat by defending its sides from the cold air of the atmosphere, is 8½ inches high and just 11 inches in diameter.

When this boiler is used for preparing only one dish of victuals, or for cooking several things that may, without inconvenience, be all boiled together in the same water, it may be covered with the cover represented in the following figure : —

Fig. 79.

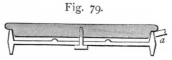

This cover is composed of one piece of cast iron, covered above with a flat circular piece of wood which serves for confining the heat.   The wood is fastened to the iron by means of a strong wood screw, with a flat square head, which passes through a hole in the centre of the piece of cast iron.

The handle of this cover must project on one side, and must be fastened to the metal and not to the wood. A piece of it is seen (at *a*) in the figure. It may either be cast with the cover, or it may be of wrought iron and fastened to it by rivets.

The figure, which is a vertical section of the cover, shows the form of it distinctly, and it will be perceived that the piece of cast iron is of a shape which renders it easy to be moulded and cast. The two small projections on the right and left of the hole in the centre of the cover are sections of a circular projection, about $\frac{2}{10}$ of an inch in height, which, as will be seen presently, is designed to serve a particular purpose. In the circumference of this horizontal projecting ring there are three equi-distant projecting blunt points, each about $\frac{3}{10}$ of an inch high above the level of the upper flat surface of the cover, or about $\frac{1}{10}$ of an inch higher than the ring from the upper part of which they project. These three points serve for supporting a shallow dish in which vegetables or any other kind of victuals is put in order to its being cooked in steam.

*Of the Manner of using this simple Apparatus for cooking with Steam.*

This may easily be done in the following manner. The flat circular piece of wood belonging to the cover of this boiler being removed and the (cast iron) cover being put down upon the boiler, a shallow dish about 2 inches less in diameter than the cover at its brim or upper projecting rim, containing the victuals to be cooked in steam, is to be set down upon the cover, just in the centre of it; and an inverted earthen pot, or any other vessel of a form and size proper for that use, being put over it, the steam from the boiler passing up through the hole in the centre of the cover will find its way under the shallow dish, and passing upwards by the sides of this dish will enter the inverted earthen pot, and, expelling the air, will take its place, and the victuals in the dish will be surrounded on every side by hot steam.

Instead of an earthen pot, an inverted glass bell may be used for covering the victuals in the shallow dish, which will not only render the experiment more striking and more amusing, but will also in some respects be more convenient; for, as the process that is going on may be seen distinctly through the glass, a judgment may, in many cases, be formed, from the *appearance* of the victuals when they are sufficiently done, without removing this vessel by which the steam is confined.

I would not, however, recommend glass vessels for common use, as they would be too expensive for poor families and too liable to be broken. For *them*, a pot of the commonest earthen-ware, or a small wooden tub, would be much more proper. But, for those who can afford the expense and who find amusement in experiments of this kind, the glass bell will be preferable to an opaque vessel.

The manner in which this simple apparatus for cooking with steam is to be arranged will be so easily understood from what has been said, that a figure can hardly be necessary to form a clear and satisfactory idea of it. I shall therefore now proceed to a description of another method of cooking with steam with these small portable kitchen boilers.

The following figure, which is drawn to a scale of 8 inches to the inch, represents a vertical section of a steam-dish of earthen-ware, proper to be used with the boiler represented by the Fig. 78 : —

Fig. 80.

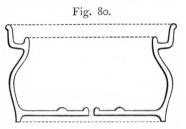

The following figure represents a vertical section of an earthen bowl, which, being inverted, may be used occasionally as a cover for the steam-dish represented above, or as a cover for the boiler : —

Fig. 81.

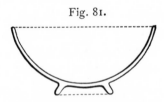

When this dish is not in use as a cover for the steam-dish or the boiler, it may be made use of for other purposes. It may, for instance, serve for bringing the soup or any other kind of food upon the table, or for containing any thing that is to be put away. In short, it may be employed for any purpose for which any other earthen bowl of the same form and dimensions would be useful.

In like manner the steam-dish may be made use of for many other purposes besides cooking with steam.

This steam-dish, and the bowl which serves as a cover to it, may both be made of cast iron ; but, when this is done, they should be tinned on the inside and japanned on the outside, to give them a neat and cleanly appearance, and prevent their rusting. They may likewise be made of pewter; or, by changing their forms a little, they may be made of tin. The choice of the material to be employed in constructing them must, in each case, be determined by circumstances.

The inverted bowl which covers the steam-dish may be used likewise for covering the boiler when the steam-dish is not in use. Or the cover of the boiler, which is represented by the Fig. 79, may be made use of instead of the inverted bowl for covering the steam-dish, and the bowl may be omitted altogether. One principal reason why I proposed this bowl was to show how by a little contrivance, an article useful in housekeeping might, without any inconvenience or impropriety, be made to serve different purposes.

It is the interest of so many persons to *increase* as much as possible the number of articles used in house-keeping, and to render them as expensive as possible, that I could not help feeling a strong desire to counteract this tendency in some measure, at least in as far as it affects the comforts and enjoyments of the poor.

The natural and the fair object of the exertions of the industrious part of mankind being the acquirement of wealth, *their* ingenuity is employed and exhausted in supplying the wants and gratifying the taste of the rich and luxurious.

It is not *their* interest to encourage the practice of economy, except it be *privately*, in their own families.

Though I sometimes speak with indignation of some of those ridiculous forms under which unmeaning and ostentatious dissipation too often insults common decency, and mortally offends every principle of good taste and elegant refinement, I am very, very far from wishing to diminish the expenses of the rich.

I well know that the free circulation of the blood is not more essentially necessary to the health of a strong athletic man than the free and *rapid* circulation of money is necessary to the prosperity of a great manufacturing and commercial country, whose power at home and abroad is necessarily maintained at a great expense.

Those who would take the trouble to meditate profoundly on the influence which taxes and luxury necessarily have, and ever must have, in promoting that circulation, would, I am confident, become more reconciled to the present state of things, and less alarmed at the progressive increase of public and private expense.

It is apathy and a general *corruption of taste* (which is inseparably connected with avarice and *a corruption of morals*), and not the progress of elegant refinement, that is a symptom of national decline.

But to return to my subject. The boiler above recommended (see Fig. 78) is peculiarly well adapted for being used with the small portable furnaces described in the *eleventh* chapter of this Essay; and, as these furnaces will not be expensive, I would strongly recommend them for the use of poor families, to be used with the utensils I have just been describing.

A cast-iron portable furnace, with one of these boilers and one of the cheap tea-kettles described in the last chapter, which might all be purchased for a small sum, would be a most valuable acquisition to a poor family. It would not only save them a great deal in fuel and in time employed in watching and keeping up the fire in cooking their victuals, but it would also have a powerful tendency to facilitate and expedite the introduction of essential improvements in their cookery, which is an object of much greater importance than is generally imagined.

The boiler in question (represented in the Fig. 78) is made double, or rather it is suspended in a hollow cylinder of sheet iron. This hollow cylinder is certainly useful, as it serves to confine the heat about the boiler; but as it renders the implement more expensive, and may wear out or be destroyed by rust after a certain time, I shall now show how a boiler, proper to be used with one of the portable furnaces before recommended, may be so constructed as to answer without a hollow cylinder.

The following figure represents a vertical section of such a boiler of cast iron drawn to a scale of 8 inches to the inch : —

Fig. 82.

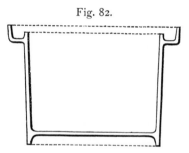

The essential difference between this boiler and that last described consists in a rim of about ¾ of an inch in depth, which descends below its bottom, and forms a kind of foot, on which it stands. This foot being made of such diameter as to fit the sand-rim of the furnace, into which it enters when the boiler is placed over the furnace, the flame and smoke of the fire are confined under the bottom of the boiler quite as effectually as if the boiler were suspended in a cylinder.

It can hardly be necessary that I should observe here — what would probably occur to the reader without my mentioning it — that stewpans and saucepans for register-stoves, and for portable furnaces of all kinds with steam-rims, might be constructed on this simple principle.

It is on this principle that the tea-kettles are constructed that were recommended in the last chapter.

I shall finish this chapter by a few observations respecting the means that may be used for combining the method of cooking here recommended for poor families, with the warming of their habitations in cold weather. This can most readily be done by using an inverted, tall, hollow, cylindrical vessel of tin, thin sheet iron, or sheet copper, as a cover to the boiler (or to the steam-dish, when that is used).

This will change the whole apparatus into a steam-stove, which, as I have elsewhere shown, is one of the best kinds of stoves that can be used for warming a room.

Whenever this is done, care must be taken to stop up the chimney fire-place with a chimney-board, otherwise all the air warmed by the stove, and rendered lighter than the external air, will find its way up the chimney, and escape out of the room. A small opening must, however, be left for the tube which carries off the smoke from the portable furnace into the chimney.

But, whenever it is intended that a portable kitchen furnace should be used occasionally for warming a room by means of steam, it will be very advisable to construct the furnace with an opening on one side of it, for the purpose of introducing the fuel without removing the boiler.

But even should no use whatever be made of this cooking apparatus in warming the room, the use of it will nevertheless be found to be very economical. The quantity of fuel consumed in preparing food will be greatly diminished; and, as a fire may at any time be lighted in one of these portable furnaces almost in an instant, there will be no longer any necessity nor any excuse for constantly keeping up a fire on the hearth in warm weather, which is but too often done in this country, even in places where fuel is neither cheap nor plenty. And even in winter, when a fire in the grate is necessary to render the room warm and comfortable, it will still be good economy to light a small separate fire in a portable furnace, or other closed fire-place, for the purpose of cooking; for nothing is so ill-judged as most of those attempts that are so frequently made by ignorant projectors *to force the same fire to perform different services at the same time.*

The *heat* generated in the combustion of fuel is a *given quantity;* and the more *directly* it is applied to the object on which it is employed, so much the better, for the less of it will escape or be lost on the way, and what is taken away on one side for a particular purpose can produce no effect whatever on the other, where it is not.

## CHAPTER XIV.

*Miscellaneous Observations respecting culinary Utensils of various Kinds, etc.—Of cheap Boilers of Tin and of cast Iron, suitable to be used with portable Furnaces. — Of earthen Boilers and Stewpans proper for the same Use. — Of* LARGE PORTABLE KITCHEN FURNACES, *with Fire-place Doors. — Description of a very cheap* SQUARE BOILER *of sheet Iron, suitable for a* PUBLIC KITCHEN. — *Of* PORTABLE BOILERS *and Fire-places that would be very useful for preparing Food for the Poor in Times of Scarcity. — Of the* ECONOMY OF HOUSE-ROOM *in the Arrangement of a Kitchen for a large Family. — A short Account of the* COTTAGE GRATE *and of a small* GRIDIRON GRATE *for open Chimney Fire-places. — A Description of a* DOUBLE DOOR *for closed Fire-places.*

ALTHOUGH my Essays are professedly *experimental*, and I seldom or never presume to trouble the public with mere speculations, or to recommend any mechanical contrivance till I have been convinced of its utility *by actual experiment*, yet my inquiries have been so numerous and so varied that I am frequently apprehensive of embarrassing my reader, and perhaps tiring and disgusting him by too great a variety of detail. To avoid that evil (which would be fatal to all my hopes) I shall, in this chapter, pass as rapidly as possible over a great number of different objects, many of which will, no doubt, be considered as curious and important. And to relieve the attention of the reader, and also to make it easy for him to pass over what he may have no curiosity to examine, I shall divide my subject as much as possible, and shall treat each distinct branch of it under a separate head of inquiry.

I shall likewise make a liberal use of figures, for by means of them it is often possible to convey more satis-factory information at a single glance than could be obtained by reading many sentences. Whenever I sit down to write, I feel my mind deeply impressed with a sense of the respect which I owe, as an individual, to the public, to whom I presume to address myself; and often consider how blamable it would be in me, especially when I am endeavouring to recommend economy, to trifle with the time of thousands.

Too much pains cannot be taken by those who write books to render their ideas clear, and their language concise and easy to be understood.

*Hours* spent by an author in saving *minutes* or even *seconds* to his readers is time well employed. But I must hasten to get forward.

*Of the Construction of cheap Boilers and Stewpans of Tin or cast Iron, proper to be used with small port-able Furnaces.*

These utensils, when they are made of tin, may be constructed on the same principles as the tea-kettles described in the last chapter; that is to say, their bot-toms being raised up about half an inch above the level of the lower part of their conical or cylindrical sides, and being moreover made of a proper diameter to fit the sand-rim of the furnace, they may be used without being made double. When they are of cast iron, they may be made of the same form below as the boiler represented by the Fig. 82, and particularly described in the last chapter.

*Of earthen Boilers and Stewpans proper to be used with portable Furnaces.*

Although the earthen stewpan represented by the Fig. 74 (see chapter XII.) is of a good form, yet those represented by the two following figures have likewise their peculiar merit. They are of forms which render them well adapted for being suspended in hollow cylinders of sheet iron, and for their being defended by those cylinders from being broken by accidental falls and blows. From a bare view of them the reader will

Fig. 83.                                    Fig. 84.

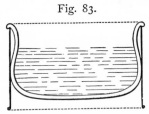 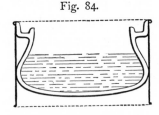

be able to appreciate their relative merit, and also to discover the particular objects had in view in the contrivance of them. The second (Fig. 84) has a steam-rim, and consequently may be used for cooking with steam by means of a steam-dish.

It would no doubt be very possible to construct earthen boilers and stewpans of such forms as to render them capable of being used with portable furnaces without being suspended in hollow cylinders. An earthen stewpan or saucepan, of the form represented by the following figure, would probably answer for that purpose : —

Fig. 85.

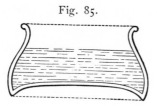

*Of large portable Kitchen Furnaces with Fire-place Doors.*

The following figure represents a vertical section (drawn to a scale of 12 inches to the inch) of a portable furnace of this kind, constructed of sheet iron: —

Fig. 86.

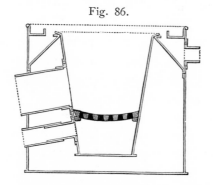

Furnaces of this kind might, I am confident, be made very useful in many cases. Wood, coals, charcoal, or turf, might indifferently be used with them; and no contrivance is better calculated for promoting both the economy of fuel and that of house-room.

Portable furnaces on this principle might easily be made of cast iron, which would be both cheap and durable; or they might be constructed partly of cast iron and partly of sheet iron, in the manner recommended in the eleventh chapter, in respect to portable furnaces without fire-place doors.

The door belonging to this fire-place is not repre-sented in the foregoing figure. It may be a hollow cylindrical stopper made of sheet iron.

### Description of a very cheap square Boiler of sheet Iron, suitable for a public Kitchen.

As some of the most wholesome and nourishing as well as most palatable kinds of food that can be pre-pared are rich and savoury soups and broths, and as many of these can be afforded at a very low price, especially when they are made in large quantities, there is no doubt but the use of them will become more gen-eral, and that they will in time constitute an essential, if not the principal, part of the victuals furnished to the poor, in every country, from public kitchens; and also to those who are lodged in hospitals or confined in prisons. And as the rich flavour and nutritious qual-ity — or, in other words, the *goodness* of any soup — depend very much on *the manner of cooking it,* — that is to say, on its being boiled or rather *simmered* for a long time over a very slow fire, — the form of the boiler and the form of the fire-place are both objects of great importance.

The simplicity and cheapness of the machinery, and the facility of procuring it in all places and getting it fitted up, are also objects to which much attention ought to be paid. Refined improvements, which require great accuracy in the execution and much care in the management of them, must not be attempted.

The boiler I would propose for the use of public kitchens is similar in all respects to that which has been adopted at Hamburg, after a model sent from Munich; for, although there is nothing about this boiler that indicates the display of much ingenuity in its contrivance, yet it has been found to answer very well as often as it has been tried; and its great simplicity renders it peculiarly well adapted for the use for which it is recommended.

A perfect idea of this boiler may be formed from the following figure, where it is represented without the wooden curb to which it is fixed when it is set in brickwork : —

Fig. 87.

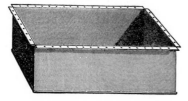

This boiler is 24 inches wide, 36 inches long, and 15 inches deep; consequently, when it is filled to within 3 inches of its brim, or when the liquor in it stands at the depth of 12 inches, it contains 10,364 cubic inches, which make above 36½ beer-gallons.

It should be constructed of sheet iron tinned on the inside; and, when it is not in use, care should be taken to wipe it out very dry with a dry cloth to prevent its being injured by rust; and, as often as it is put away for any considerable time, it should be smeared over with fresh butter or any other kind of animal fat unmixed with salt.

The sheet iron will be sufficiently thick and strong if the boiler when finished weigh 40 pounds; and, as the best sheet iron costs no more than about 3½*d*. per lb., the manufacturer ought not to charge more than 6*d*. per lb. for the boiler when finished, which, if it weigh 40 lbs., will amount to 20*s*.

To strengthen the boiler at the brim, it must be fastened to a curb of wood, which may be a frame of board $1\frac{1}{4}$ or $1\frac{1}{8}$ inch thick, 5 inches wide, and just large enough to allow the boiler to pass into it and be suspended by its projecting brim. This brim, which may be made about an inch wide, must be fastened down upon the wooden curb with tinned nails or with small wood screws.

This curb will be 3 feet 10 inches long and 2 feet 10 inches wide; and, as the stuff used is 5 inches wide, it will measure very nearly $2\frac{3}{4}$ feet, superficial measure, which, at 6*d.* the foot (which would be a fair price in London for the work when done), would amount to $1s.\ 4\frac{1}{2}d.$

The boiler must be furnished with a cover, which may be made of wood, and should consist of three distinct pieces framed and panelled, and united by two pair of hinges as they are represented in the following figure:—

Fig. 88.

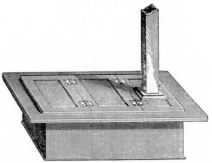

This cover will measure about 7 superficial feet, and, at 7*d.* the foot, will cost 4*s.* 1*d.* The hinges may cost about 4*d.* the pair, consequently the cover will cost, all together, about 4*s.* 9*d.*

This figure represents the boiler fixed in its wooden curb and with its cover in its place.

The first division of the cover (which is 12 inches wide) is laid back on the second (which is 14 inches wide) whenever it is necessary to open the boiler to put anything into it or to take anything out of it, or merely to stir about its contents. When the boiler is to be washed out and cleaned, the opening into it is made larger by throwing back the first and second divisions of its cover, folded one upon the other, and leaning them against the steam-tube which stands upon the third division of the cover, which division is firmly fixed down upon the curb of the boiler by means of wood screws.

The steam-tube (which should be of sufficient length to carry the steam from the boiler out of the room into the open air or into a neighbouring chimney) may be made of four slips of $\frac{3}{4}$ inch thick deal boards fastened together (by being grooved into each other and nailed together) in such a manner as to form a hollow square trunk, measuring about $1\frac{1}{4}$ inches wide in the clear.

In setting this boiler in brick-work, the flame and smoke from the fire should be made to act on its bottom only, but its sides and ends should be bricked up, in order more effectually to confine the heat. The mass of brick-work should be just 3 feet 8 inches long and 2 feet 8 inches wide, in order that the curb of the boiler may cover it above and project beyond it horizontally on every side about $\frac{1}{2}$ an inch. The bars of the fire-place on which the fuel burns should be situated 12 or 14 inches below the bottom of the boiler, in order that the boiler may not be injured when the fire happens by accident or by mismanagement to be made too intense.

It is not necessary that I should mention here any of the precautions which are to be observed in setting boilers of this kind in brick-work; for that subject has already been so amply treated in various parts of these Essays that to add any thing to what has already been said upon it could be little better than an unnecessary and tiresome repetition.

This boiler would be sufficiently large for cooking for about 300 persons. If it were necessary to feed a much greater number from the same kitchen, I would rather recommend the fitting up of two or more boilers of this size than constructing one large boiler to supply the place of a greater number of others of a moderate size; for I have found by much experience that very large boilers are far from being either economical or convenient.

Large boilers of sheet iron, and especially such as are not kept in constant use, are always *very expensive*, on account of their being so liable to be destroyed by rust.

*Of portable Boilers and Fire-places that would be very useful for preparing Food for the Poor in Times of Scarcity.*

There is always much trouble and inconvenience, and frequently much danger, in collecting together great numbers of idle people; and these assemblies are never so likely to produce mischievous effects as in times of public calamity, when it is peculiarly difficult to preserve order and subordination among the lower and most needy classes of society.

I have often trembled at seeing the immense crowds of poor people, without occupation, who were sometimes collected together at the doors of the great public kitchens in London during the scarcity of the year 1800.

Two or three hundred people may, without any considerable inconvenience, be supplied with food from the same kitchen; but when public kitchens are not connected with asylums or houses or schools of industry where the poor assemble to work during the day, and when there is no other object in view but merely to enable the poor to purchase good and wholesome food at the lowest prices possible, without any interference at all with their domestic employments or concerns, it appears to me that it would always be best to select from amongst the poor a certain number of honest and intelligent persons, and encourage them to prepare and sell to their poor neighbours, under proper regulation and inspection, such kinds of food and at such prices as should be prescribed by those who have the charge of providing for the relief of the poor.

A plan of this sort might be executed at any time on the pressure of the moment, without the smallest delay, and almost without either trouble or expense, if each parish or community were to provide and keep ready in store a certain number of portable kitchen furnaces, with boilers belonging to them, to be lent out occasionally to those who should be willing to undertake to cook and sell victuals to the poor on the terms that should be proposed.

If these boilers were made to hold from 8 to 10 gallons, they would serve for preparing food for 60 or 70 persons: and, as they would require very little fuel, and so little attendance that a woman who should undertake the management of one of them might perform that service with great ease by devoting to it each day the labour of half an hour, and giving to it occasionally a few moments of attention, which would hardly interrupt her in her common domestic employments, this method of preparing food would be very economical, — perhaps more so than any other, — and, with proper inspection, it would be little liable to abuse.

How very useful would these portable boilers and furnaces be for providing a warm and cheap dinner for children who frequent schools of industry!

No furnace could, in my opinion, be better contrived for this use than that represented in the Fig. 86; and the boiler might be made either of sheet iron tinned, or of copper tinned, or of cast iron. It cannot be necessary that I should give any particular directions respecting its form, and its dimensions may easily be computed from its capacity, when that is determined on.

A portable cooking apparatus of this kind, which is designed as a model for imitation, may be seen in the repository of the Royal Institution.

## Of the Economy of House-room in the Arrangement of a Kitchen for a large Family.

There is nothing which marks the progress of civil society more strongly than the use that is made of house-room; and nothing would tend more to prevent the too rapid progress of destructive luxury among the industrious classes than a taste for neatness and true elegance in all the inferior details of domestic arrangement. The pleasing occupation which those objects of rational pursuit afford to the mind fills up leisure time in a manner that is both useful and satisfactory and prevents *ennui* and all its fatal consequences.

The poor cook their victuals in the rooms in which they dwell; but those who can afford the expense — and many indeed who cannot — set apart a room for the purpose of cooking, and call it a kitchen. I am far from desiring to alter this order of things, for I think it perfectly proper. What I wish is, that each class of society may be made as comfortable as possible, and that all their domestic arrangements may be *neat* and *elegant*, and at the same time *economical.*

I always fancy that teaching industrious people economy, and giving them a taste for the improvement of all those useful contrivances and rational enjoyments that are within their reach, is something like showing them how, without either toil or trouble, and with a good conscience, they may obtain all those advantages which riches command, together with many other very sweet enjoyments which money cannot buy. And whose heart is so cold as not to glow with ardent zeal at a prospect so well calculated to awaken all the most generous feelings of humanity?

But to return from this digression. There are various methods that may be used for economizing house-room in making the necessary arrangements for cooking. If the family be small, the use of portable furnaces and boilers will be found to be very advantageous.

For a large family I would recommend what I shall call a *concealed kitchen.* There are two very complete kitchens of this kind, which have been fitted up under my direction at the Royal Institution: the one, which is small, is in the housekeeper's room; the other is in the great kitchen. These were both made as models for imitation, and may be examined by any person who wishes to see them.

There are also two kitchens of this kind in my house at Brompton in two adjoining rooms, which have been fitted up principally with a view to showing that all the different processes of cookery *may* be carried on in a room which, on entering it, nobody would suspect to be a kitchen. The following figure is the ground plan of one of them: —

Fig. 89.

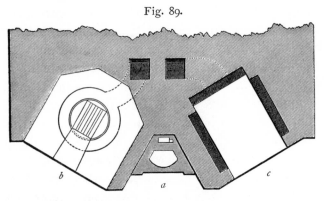

*a* is the opening of the fire-place, which is brought forward into the room about 14½ inches. This was done, in order to give more room for the family boiler, which is situated at *b*, and the roaster, which is placed on the other side of the open chimney fire-place at *c*.

The two broad spaces on the two sides of the roaster, by which the smoke from the fire below it rises up round it, and another at the farther end of it, by which the smoke descends, are distinguished by dark shades, as are also the two square canals by which the smoke from the roaster and that from the boiler rise up into the chimney.

The top of the grate is seen which belongs to the open chimney fire-place: it is represented by horizontal lines. It is what I have called a *cottage grate*, and what is sold in the shops under that name. The retail price of this grate, with its fender and trivet, is *ten shillings and sixpence.* The Carron Company entered into an engagement with me to furnish them by whole-sale to the trade, delivered in London, at *seven shillings and sixpence.* A front view of this grate may be seen in the next figure. As this figure (Fig. 89) is designed merely for showing *where* the different parts of the apparatus are to be placed, and not *how* they are to be fitted up, none of the details of the setting of the roaster or boiler were in this place attempted to be expressed with accuracy. Information respecting those particulars must be collected from other parts of the work.

The grate represented in this figure is calculated for boiling a pot or a tea-kettle, and for heating flat-irons for ironing. Its bottom is so contrived as to be easily taken away and replaced. By removing it at night, or whenever a fire is no longer wanted, the coals in the grate fall down on the hearth, and the fire immediately goes out. This contrivance not only saves much fuel, which otherwise would be consumed to waste, but it is also very convenient on another account. As all the coals and ashes fall out of the grate when its bottom is removed, on replacing it again the grate is empty and ready for a new fire to be kindled in it.

The top of this grate, which is a flat piece of cast iron, has one large hole in it for allowing the smoke to pass upwards, and another behind it, which is much smaller, through which it is forced to *descend* into what has been called a *diving-flue,* whenever the boiler be-longing to this fire-place is used, — which boiler is suspended in a hollow cylinder of sheet iron, about

$11\frac{1}{2}$ inches in diameter, resembling in all respects the boilers used with the register-stoves described in the tenth chapter of this Essay.

I intend, as soon as it shall be in my power, to publish a particular detailed account of this grate, and also of several others for open chimney fire-places, which at my recommendation have lately been introduced in this country. In the mean time, I avail myself of this opportunity of pointing out one fault which has been committed by almost all those who have undertaken to set *cottage grates* in brick-work. They have made what has been called the *diving-flue* much too deep. It is more than probable that the name given to this flue has contributed not a little to lead them into this error. When properly constructed, it hardly deserves the name of *a flue*, for it ought not to be above *two inches deep*, measured from the under surface of the flat plate of cast iron which forms the top of the grate. There are two important advantages that result from making this opening in the brick-work for the passage of the smoke *very shallow:* the one is, that in this case it may easily be cleaned out when coals happen to fall into it by accident when it is left uncovered; and the other is, that the back wall of the fire-place, against which the fuel burns, may in that case be made thick and strong, and not so liable to be destroyed by the end of the poker in stirring the fire as it is when there is a hollow flue just behind it.

Both these are important objects, and for want of due attention being paid to them cottage grates have, to my knowledge, often been disgraced and rejected. When they are properly set and properly managed, they are very useful fire-places where coal or turf is burned; and it never was designed that they should be used with wood.

When kitchens are fitted up on the plan here recommended in places where wood is used as fuel, the open chimney fire-place, which is situated between the roaster and the boiler, may be constructed *of the form* represented in the foregoing figure, but without any fixed grate; and the wood may be burned on andirons or on a small movable *gridiron grate* placed on the hearth.

These *gridiron grates* are very simple in their construction, cheap and durable; and they make an excellent fire, either with coals or turf, or with wood, if it be sawed or cut into short billets. Five of these grates may be seen at the house of the Royal Institution: one in the great lecture-room, one in the apparatus-room, one in the manager's room, one in the clerks' room, and one in the dining-room. They have hitherto been made of two sizes only; namely, of 16 inches and of 18 inches in width in front. The width of the back part of the grate is always made just equal to half its width in front, and the two sloping sides or ends of the grate are each just equal in width to the back. The form and dimensions of the grate determine the form and dimensions of the open chimney fire-place in which it is used; for the back of the fire-place must always be made just equal in width to the back of the grate, and the sloping of the covings must be the same as the sloping of the ends of the grate.

From what has been said of the proportions of the front, back, and sides of these grates, it is evident that the covings and backs of their fire-places must make an angle with each other just equal to 120 degrees. This angle I have been induced to prefer to one of

135 degrees, which I formerly recommended for open chimney fire-places. The reasons for this preference will be fully explained in another place. To give them here would take up too much time, and would moreover be foreign to my present subject.

For the information of the public, and to prevent, in as far as it is in my power, exorbitant demands being made for these useful articles, I would just observe that the smallest or 16-inch *gridiron grate*, together with all the apparatus belonging to it, ought to cost, *by retail*, no more than *seven shillings*. This apparatus consists of a cast-iron fender, a trivet for supporting a boiler or a tea-kettle over the fire, and a small plate of cast iron (to be fastened into the back of the chimney), by means of which, and a small bolt or nail, the grate is fastened in its place on the hearth.

The second-sized or 18-inch *gridiron grate*, with all its apparatus (consisting of the three articles mentioned above), ought to be sold, by retail, for *seven shillings and sixpence*.

The *wholesale price* of these articles, at the Carron Company's warehouse, in London (Thames Street, near Blackfriars' Bridge), to the trade, and to gentlemen who buy them by the dozen, to distribute them to the poor, is : —

> For the gridiron-grate  No. 1, with
>   the articles belonging to it.  .  .  *four shillings.*
> For that  No. 2, with  the articles
>   belonging to it.  .  .  .  .  .  .  *four shillings and sixpence.*

These are the wholesale and retail prices which I fixed with the agent of the Carron Company, at their works in Scotland, in the autumn of the year 1800, when I made a journey there for the purpose of estab-

lishing these regulations; and when I made a present
to the Company of all my patterns, which I had got
made in London, and which had been rendered as perfect as possible by previous experiments, — namely, by
getting castings taken from them by the best London
founders, and altering them occasionally, till they were
acknowledged to be quite complete.

If it had been possible for me to have done more to
prevent impositions, I should have done it with pleasure; and I should have felt, at the same time, that I
had done no more than what it was my duty to do.

But to return from this long digression. I shall now
hasten to finish my account of the means which have
been used in one of the rooms in my house (that destined for the large kitchen) for concealing the roaster
and the family boiler.

The following figure is an elevation of that part
of the side of the room where these implements are
concealed : —

Fig. 90.

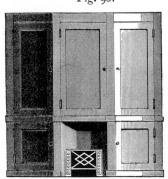

The open chimney fire-place and the front of the
grate are distinctly shown in the middle of this figure,
in the lower part of it. The panelled door, immedi-

ately above the mantel of the chimney fire-place, which
reaches nearly to the ceiling of the room, serves to
shut up a small closet with narrow shelves, which has
no connection with culinary affairs, but is used for
putting away candlesticks, and any other small articles
used in housekeeping, which are occasionally laid by
when not in actual use. The two other panelled doors
by the side of it serve, — the one (that on the right
hand) for concealing the roaster, and the other for
concealing the family boiler.

The two (shorter) panelled doors, on the right and
left of the open chimney fire-place, and on the same
level with it, serve for concealing the fire-place doors
and ash-pit doors of the closed fire-places of the roaster
and of the boiler.

The steam from the boiler (after passing through the
steam-dishes, when they are used) is carried off by a
tin tube into a small canal, which conveys it into the
chimney in such a manner that no part of it comes
into the room. The steam from the roaster is carried
off in like manner by its steam-tube.

If a void space, about 2 or 3 inches in depth, be left
between the outside of the door of the roaster and the
inside of the panelled door which shuts it up and con-
ceals it, and if this panelled door be lined on the inside
with thin sheet iron, the process of roasting may be
carried on with perfect safety with this door shut. And
if similar precautions be used to defend the other panelled doors from the heat, they may also be kept shut
while the processes of boiling and roasting are actually
going on.

By these means it would be *possible* to prepare a
dinner for a large company in a room where there should

be no appearance of any cooking going on. But I lay no stress on this particular advantage resulting from this arrangement of the culinary apparatus. The real advantage gained by it is this : that the kitchen is left an *habitable*, and even an *elegant room*, when the business of cooking is over.

The kitchen in Heriot's Hospital at Edinburgh, which was fitted up in the autumn of the year 1800, is arranged in this manner, — with this difference, however, that all the panelled doors are omitted. The boiler is shut up by a door of sheet iron, japanned ; and the door of the roaster and the two fire-place doors and two ash-pit register doors are exposed to view.

As the brick-work is whitewashed and kept clean, and as the doors are all either japanned black or kept very clean, the whole has a neat appearance.

The roaster and principal boiler in the great kitchen of the house of the Royal Institution are put up nearly in the same manner as those in Heriot's Hospital, excepting that in the former there is a hot closet, which is situated immediately above the roaster, whereas there is none belonging to the latter.

In one of the kitchens in my house there is, in the place of the roaster, a roasting-oven, with a common iron oven of the same dimensions placed directly over it, and heated by the same fire.

The door of my roaster and that of my roasting-oven are made single, of thin sheet iron, and they are covered on the outside with panels of wood, for confining the heat. Instead of doors to their closed fire-places, I use square stoppers, made of fire-stone or hard fire-brick, fastened to flat pieces of sheet iron, to which knobs of wood are fixed, which serve instead of handles.

These stoppers answer for confining the heat quite as well, and perhaps even better, than double doors, and they cost much less. They are fitted into square frames of cast iron (nearly similar to that represented in the Fig. 91), which are firmly fixed in the brick-work by means of projecting flanges, which are cast with them. The front edge of this frame or doorway is ground and made perfectly level; and the plate of sheet iron, which forms a part of the stopper, being made quite flat, shuts against the front edge of this doorway, and closes the entrance into the fire-place with the greatest accuracy.

The entrance into the ash-pit is likewise closed by a stopper, which is so contrived as to serve occasionally as a register for regulating the quantity of air admitted into the fire-place.

As this *register-stopper* for the ash-pit of a small closed fire-place is very simple in its construction, and as I have found it to answer very well the purpose for which it was contrived, I shall present the reader with the following sketch of it, which will, I trust, be sufficient to enable a workman of common inge-

Fig. 91.

nuity to construct, without difficulty, the thing which is represented.

The box with a flange at each of its ends forms the

door-way into the ash-pit. It is of cast iron, and its opening in front is 7¼ inches wide and 3¾ inches high. It is concealed in the brick-work in such a manner that its front edge only is seen, projecting about ⅛ of an inch before the brick-work.

When the register-stopper belonging to this door-way (which is shown in this figure) is pushed quite home, its flat plate comes into contact with the front edge of the door-way, and closes the passage into the ash-pit so completely that no air can enter. By withdrawing this stopper more or less, more or less air is admitted. The narrow, thin, elastic bands of iron, the ends of which are fastened by rivets to the flat plate of the stopper, serve to confine the stopper in any situation in which it is placed, which service they are enabled to perform (in consequence of their elasticity and of their peculiar shape) by pressing against the sides of the door-way.

The only objection that I am acquainted with to this kind of register for the door-way of the ash-pit of a small closed fire-place is that it is not quite so easy to see the precise state of the register as it is when the air is admitted through a hole in the front of the ash-pit door in the usual manner; but this objection is of no great importance, especially as means may easily be devised to remedy that trifling defect.

The door-way frames to all the closed fire-places in my own kitchen are in all respects like that represented in the foregoing figure (Fig. 91), with this difference only, that they are 5 inches high instead of being 3¾ inches in height. An account has already been given of the manner in which their stoppers were constructed.

It is right that the reader should be informed that although I have made use of stoppers to close the passage into each of the closed fire-places in my own kitchen, yet very few persons have adopted this simple and cheap contrivance. The reason why it has not come into more general use might easily be explained; but I fancy it will be best that I should say nothing now on that subject. Instead of recommending what nobody would find much advantage in furnishing at a fair price, it will be more wise and prudent to give a short description of a more complicated, more elegant, and more expensive contrivance, which has already found its way into the shops of several of the most respectable ironmongers in London. As this contrivance has often been used, and has always been found to answer perfectly well, I can venture to recommend it to all those to whom an additional expense of a few shillings or a guinea or two in fitting up a kitchen is not considered as an object of importance.

### *A short Description of a* DOUBLE DOOR *for a closed Fire-place.*

The following figure (which is drawn to a scale of 6 inches to the inch) represents a horizontal section of one of these double doors, and also of a part of the brick-work in which it is set.

A is the inside door, and B is the outside door. These doors are so connected by means of a crooked rod of iron *f*, and the two joints *g* and *h*, that when the outside door is opened or shut the inside door is necessarily opened or shut at the same time. The inside door, which is of cast iron and near $\frac{1}{2}$ an inch in thick-

ness, is movable on two pivots, one of which is represented at *e.*   The outside door is movable on two
hinges, one of which is shown at *d.*

*c* is the latch by which the outside door is fastened.
This is of such a form that it may be used as a latch,
and may serve at the same time as a handle for opening and shutting the door.

Fig. 92.

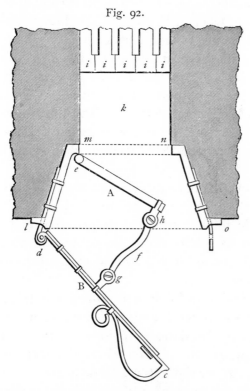

The door-way, which is of cast iron, is in the shape
of a hollow truncated quadrangular pyramid, with a
flange in front, about an inch wide, which flange,
when seen in front, seems to form a kind of frame to
the outside door; the flange, which is about $\frac{1}{4}$ of an inch

in thickness, projecting before the vertical front of the brick-work.

*l, m, n, o*, represents a horizontal section of this cast iron door-way. The brick-work in which it is set is distinguished by diagonal lines.

*k* is the passage leading to the fire-place: it is 6 inches wide in the clear from *m* to *n*, 5 inches high, and 6 inches long, measured from the inside of the inside door, when it is shut, to the hither ends of the openings between the iron bars of the fire-place, through which openings the air comes up from the ash-pit into the fire-place. The hither ends of these bars (five in number) are represented in the figure. They are each distinguished by the letter *i*. The opening of the in-side door-way is 6 inches wide and 5 inches high in the clear; and the door itself is $6\frac{1}{2}$ inches wide and $5\frac{1}{2}$ inches high.

The outside door-way is 10 inches wide and 9 inches high in the clear; and the door, which is about $\frac{2}{10}$ of an inch in thickness, is $10\frac{1}{2}$ inches wide and $9\frac{1}{2}$ high. The extreme width of the door-frame to the outward edge of the flange is $12\frac{1}{2}$ inches, and its extreme height is $11\frac{1}{2}$ inches.

The two straps of iron to which the hooks of the hinges of the outside door are fastened pass through two holes in the flange, provided for them in casting the door-way, and are riveted to the sloping side of the door-way on the left-hand side of it.

These holes are each $\frac{7}{8}$ of an inch in length from top to bottom, and about $\frac{1}{4}$ of an inch in width. There is another similar hole in the flange on the opposite side of the door-way, through which a strap of iron passes, the end of which projecting forward before the level of the front edge of the door-way serves as a catch or hook, into which the latch of the door falls when the door is closed.

These three holes in the side flanges of the door-way are distinctly represented in the following figure, which is an elevation or front view of this door-way, without its doors: —

Fig. 93.

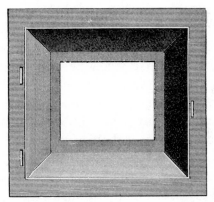

It appears by this figure, but still more distinctly by the last (Fig. 92), that the flange or front of this door-way is not quite flat. It is raised at its inward edge, which projects forward about ¼ of an inch. This projecting rim, which is cast as thin as possible, is ground upon a flat sand-stone and made quite level, in order that the outside door, which is flat, by shutting against the front of this projecting edge may close the opening into the fire-place with the greatest possible accuracy.

It will likewise be remarked, on examining this figure (Fig. 93) with attention, that the opening which is closed by the inside door is not precisely in the middle of the vertical flat surface against which that door shuts, being situated a little above the middle of it. This particular arrangement has been found to be of considerable use, as it serves to prevent small pieces of coal from getting between the inside door and that flat surface when the door is shut.

These double doors (of a size larger than that represented by the two preceding figures) have lately been introduced in a considerable number of hothouses in the neighbourhood of London; and I have been told, by several persons who have tried them, that they have been found very useful indeed. I was lately assured by a very respectable gardener, who has adopted them in all his hothouses, that since he has used them and the register ash-pit doors which belong to them and are always sold with them, and since he has altered the construction of his fire-places, his consumption of coals has been little more than half as much as it used formerly to be.

In setting these double doors in brick-work, great care should always be taken to make the entrance into the fire-place of some considerable length, or to keep the hither ends of the iron bars on which the fuel burns at some distance from the inside door; otherwise, if the burning fuel be near that door, it will heat it and its frame red-hot, which will soon destroy their form and prevent the door from closing the entrance of the fire-place with accuracy.

I have found it to be a good general rule to place the hither ends of the bars, which form the grate of the fire-place, as far beyond the inside door as that door-way is wide in the clear. And it will be found to be an excellent precaution to defend the door from the heat, if that part of the passage into the fire-place which lies beyond the inside door be kept constantly rammed quite full of small coals; or, what would be still better, of coal-dust mixed up with a certain proportion of moist clay.

I have already, in a former part of this Essay, mentioned how necessary it is, in setting double doors in brick-work, to take care to mask the farther end of the door-way in such a manner (by means of bricks interposed before it, or between it and the fire) that the rays from the burning fuel may never fall on it. The manner in which this is to be done is clearly represented in the Fig. 92.

All these precautions for preventing these double doors from being injured by excessive heat will be the more necessary in proportion as the fire-places are larger to which they belong.

There is one essential part of this apparatus which, for want of room, was omitted in the two last figures, — that is, the straps of wrought·iron, by means of which the door-way is firmly fixed in the brick-work; but this omission can be of no consequence, as every common artificer will know, without any particular directions, how that part of the work should be executed. These straps must of course be fastened to the cast-iron door-way by means of rivets.

----

## CHAPTER XV.

*Apology for the great Length of this Essay. — Regret of the Author that he has not been able to publish Plans and Descriptions of the various culinary Inventions that have lately been put up in the Kitchen belonging to the House of the Royal Institution and*

*in the Kitchen of Heriot's Hospital at Edinburgh. —*
*A short Account of a* BOILER, *on a new Construction,*
*lately put up at the House of the Royal Institution,*
*for the purpose of* GENERATING STEAM *for warming*
*the Great Lecture-Room. — This Boiler would prob-*
*ably be found very useful for* STEAM-ENGINES. —
*An Account of a Contrivance for preventing metallic*
STEAM-TUBES *from being injured by the alternate*
*Expansion and Contraction of the Metal by Heat*
*and Cold. — An Account of a simple Contrivance*
*which serves as a Substitute for* SAFETY-VALVES.

I CANNOT finish this Essay without apologizing
for the great length of it.  I had no idea when I
began it that it would ever have grown to such a volu-
minous size; but I am not conscious of having inserted
any thing that could well have been omitted.

I was very desirous of laying before the public complete plans and descriptions of the various culinary
inventions that have lately been put up in the great
kitchen of the house of the Royal Institution in Al-
bemarle Street, and also of those erected in Heriot's
Hospital at Edinburgh, in the autumn of the year
1800; but my stay in this country will be too short
for me to undertake so considerable a work at this
time.  I am happy, however, that these new contrivances, some of which have already been proved to be
very useful, are situated in places of public resort where
persons desirous of examining them may at all times
obtain free admission.

There are also several other new and useful contrivances at the house of the Royal Institution, which I
should have had great pleasure in laying before the

public, had it been in my power, as I am persuaded that correct accounts of them would have been very acceptable to men of science, and to all those who take pleasure in promoting new and useful mechanical improvements.

I should, in particular, have been very glad to have given plans and descriptions of all the various parts of the steam-apparatus that has been put up for the purpose of warming the great lecture-room. The boilers for generating the steam are, if I am not much mistaken, well worthy of the attention of those who make use of steam-engines; and as the subject is of infinite importance in this great manufacturing country, where the numerous advantages which result from the use of machinery are known and every day more and more felt by individuals and by the public, I cannot resist the strong inclination which I feel, to attempt in a few words to give a general idea of this contrivance. Those who wish to know more of the matter may get all the information respecting it which they can want by applying at the house of the Royal Institution.

*A short Account of the* Boilers *lately put up at the House of the Royal Institution for* GENERATING STEAM *for warming the Great Lecture-Room.*

Over an oblong closed fire-place, furnished with double doors, ash-pit register door, etc., are placed two cylinders of copper, laid down horizontally by the side of each other over the fire, each cylinder being 15 inches in diameter and 48 inches long. Immediately over these two cylinders, and resting on them, are placed two other cylinders of copper of the same length and diameter; and over these last, and resting on them, are placed two other like cylinders, making six cylinders in the whole, all made of the same material and being of the same dimensions.

The fire-place being situated under the hither ends of the two lower cylinders, the flame runs along under them to their farther ends, where it passes upwards and comes forward between the upper sides of the two lower cylinders, and the lower sides of the two cylinders immediately above them. Being arrived at the front wall of the brick-work, it there rises up again, and then passes along horizontally between the two middle cylinders and the two upper cylinders, till it comes to the back wall; and, passing up by the farther ends of the upper cylinders, it comes forwards horizontally, for the last time, in an arch or vault of brick-work which covers the two upper cylinders. Being arrived once more at the front wall of the brick-work, it there enters a canal (furnished with a good damper) by which it goes off into a neighbouring chimney.

These cylinders are confined in their places by being placed in pairs, over each other, between two parallel vertical walls, which are built just so far asunder as to admit two cylinders, placed horizontally by the sides of each other; and the flame is prevented from finding its way upwards between the two cylinders which lie by the sides of each other, or between the outsides of those cylinders and the sides of the vertical walls with which they are in contact, by filling up the joining between them with good clay, mixed with small pieces of fire-bricks.

The farther ends of all the cylinders are closed up, and all the tubes which are necessary for the admission of water and for the passage of the steam are fixed to a circular plate of metal, which closes (by means of flanges and screws) the front ends of the cylinders.

In consequence of this particular arrangement it will be perfectly easy to make all the cylinders of *cast iron*, even when these boilers are destined for steam-engines of the largest dimensions. The number of sets of cylindrical boilers, which in each case it will be necessary to put up, must be determined by the size of the cylinders and by the quantity of steam that will be wanted. Six cylindrical boilers put up in a separate mass of brick-work, in the manner above described, I call *one set*.

It will always be found to be very advantageous to have at least three or four sets of cylindrical boilers to each steam-engine, instead of having one set of larger cylinders; and this not only on account of the wear and tear of small fire-places being incomparably less expensive than in those which are large, but also on account of the economy of fuel which will be derived from that arrangement, and the great convenience that will be found to result from the use of small boilers, which may at any time be heated and made to boil in a very few minutes; and from the advantage of being able at all times to regulate the number of sets of boilers in use to the load on the engine.

It is quite impossible to make a small fire in a large fire-place without a great loss of heat; but, by having a number of small separate fire-places, an engine may be made to work with a light load with almost as small a proportion of fuel as when it is made to perform its full work. But to return to our cylindrical boilers.

The two lower cylinders, and those two which lie immediately over them, being destined for the generation of steam, are kept constantly about half full of water, which water they receive, already hot, from the two upper cylinders, in which last the water should never boil.

These upper cylinders communicate, by an open pipe, with a reservoir of water, which is situated several feet above them; consequently, as fast as they furnish water to the four cylinders which lie below them, that water so furnished is immediately replaced by water which comes from the reservoir above.

As the pipe which brings this water from the reservoir enters the cylinders some considerable distance below their centres, and as the pipes which convey the water from them to the cylinders below are fixed in their centres, as cold water is heavier than warm water, it is evident that the water which enters them cold from the reservoir will take its place at the lower parts of these cylinders, while only the lighter hot water will be furnished to the cylindrical boilers below.

The method of regulating the admission of water into the boilers below, where the steam is generated, is so well known that it would be superfluous to give a particular account of it.

In the set of boilers that has been put up at the house of the Royal Institution, the open ends of all the cylinders are on one side; that is to say, they all come through the front wall of the brick-work. This arrangement was rendered necessary in that particular case by local circumstances: it would, however, have been better if only the lower and upper pairs of cylinders had come through the front wall, and the open ends of the middle pair had passed through the back wall; for in that case it would have been easier to provide a passage for the flame round the ends of the middle cylinders.

One evident advantage that will be derived from constructing steam-engine boilers on the principles here recommended is their superior strength to resist the efforts of the steam, which will render it possible to use very thin sheet copper or sheet iron in constructing them, when they are made of those materials. Another advantage will be the great facility of removing and repairing any of the cylinders which may happen to leak, or which may be found to be damaged or worn out. When several sets of cylinders are put up for the same engine (which I would always recommend, even for engines of the smallest size), any of these occasional repairs may be made without stopping the engine.

If these cylindrical steam boilers should be found to be useful for steam-engines, they cannot fail to be equally so for generating steam for heating dyers' coppers by means of steam, for bleaching by means of steam, and, in general, for every purpose where steam is wanted in large quantities.

They must, I think, be peculiarly well adapted for dyers ; for, as water less hot than boiling water is frequently wanted by them in the course of their business, the upper cylinders will at all times afford a plentiful supply of warm water, which may, without the smallest inconvenience, be drawn off whenever it is wanted.

To prevent in the most effectual manner the loss of heat which is occasioned by the passage of steam through the safety-valve, that steam which so escapes out of the boiler may be carried off in a tube provided for that purpose, and conducted into the upper cylin-

ders or into the reservoir which feeds them. In doing
this, care must be taken to cause the steam to descend
perpendicularly, from the height of eight or ten feet,
before it enters the water where it is intended that it
should be condensed; and the end of the tube through
which the steam descends and enters the water should
be plunged to a certain depth below the surface of the
water.

I shall finish this chapter and conclude this Essay by
giving a short description of two very simple contrivances, which have been put in practice at the house of
the Royal Institution, and which have been found to be
very useful. The one is a contrivance for preventing
most effectually the bad effects of the alternate expansion and contraction by heat and cold of the metallic
tubes which are used in conveying steam to a considerable distance; and the other is a substitute for safety-
valves in an apparatus for heating rooms by means of
steam.

*Of the Means that may be used for preventing metallic*
   *Steam-tubes, of considerable Length, from being in-*
   *jured by the alternate Expansion and Contraction of*
   *the Metal by the different Degrees of Heat and Cold*
   *to which those Tubes are occasionally exposed.*

We will suppose the tube in question to be of copper,
and eight inches in diameter (which is the size of that
used for warming the great lecture-room at the Royal
Institution). Let this tube be made in lengths of ten
feet; and instead of joining the ends of these tubes
together immediately, to form one long tube, let a very
short tube or cylinder, of only one or two inches in
length and 24 inches in diameter, closed at each end

with a flat circular plate of sheet copper, like the head
of a drum, be interposed between their joinings. These
two circular sheets of copper, which form two ends of
this very short cylinder, must be perforated in their
centres with holes 8 inches in diameter, to give a passage to the steam; and the ends of the tubes must be
firmly fastened to them by means of flanges and rivets.

The following figure, which represents an outline of
a portion of a steam-tube constructed in this manner,
will give a clear idea of this contrivance: —

Fig. 94.

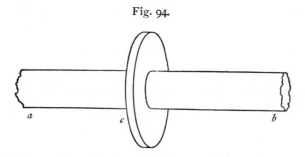

*a*, *b*, are portions of two of the tubes which are united
together by means of the short flat cylinder *c*.

Now if we suppose one of these tubes (10 feet long)
to be immovably fixed *in the middle of its length* to
a beam of wood or to a solid wall, the increase or diminution of the length of each half of it — arising from its
being occasionally heated to the temperature of boiling
water by steam, or cooled to the mean temperature of
the air of the atmosphere, — being free will cause its two
ends to push inwards or to draw outwards the two flat
ends of the two neighbouring short cylinders to which
they are attached; and, as these short cylinders are
24 inches in diameter, while the tube is only 8 inches
in diameter, the elasticity of the large circular thin

plates of metal will allow it to be pressed inwards or drawn outwards without injury, much more than will be necessary in order to give room for the expansions and contractions of the tubes.

Hence it appears that, by this simple contrivance, steam may be conveyed to any distance, however great, in closed metallic tubes, without any danger of injury to the tubes from the expansions and contractions of the metal.

*A short Description of a Contrivance which serves in-stead of Safety-valves for a Steam Apparatus, which is used for heating the Great Lecture-Room at the House of the Royal Institution.*

The following figure, which represents a vertical section of this contrivance, will give a clear idea of it, and of the manner in which it acts: —

Fig. 95.

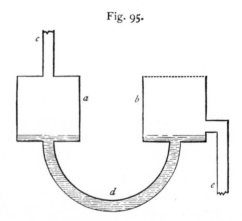

*a* and *b* are two cylinders of copper, 6 inches in diame-ter and 6 inches in length, placed in an erect position. The cylinder *a* is closed both above and below; the cylinder *b* is closed below, but is open above.

The semi-circular tube *d*, which is represented filled with water, serves to connect the two cylinders together.

By the tube *c*, the water, which results from the condensation of the steam in the steam-tubes which warm the room, returns to the reservoir which feeds the boiler. This water, after falling into the cylinder *a*, passes through the semi-circular tube *d* into the cylinder *b*, and then goes off from that cylinder, and is conveyed, still warm, to the reservoir, by the tube *e*.

This simple apparatus serves as a safety-valve in the following manner: When the steam in the steam-tubes is redundant, it descends through the tube *c*, and forcing the water out of the semi-circular tube *d* into the cylinder *b*, it follows it through that tube, and, escapes into the open air through the open end of that cylinder. When the strength of the steam is sufficiently diminished, a small quantity of water, still remaining in the lower part of the cylinder *b*, returns back into the tube *d*, and cuts off the communication between the external air and the inside of the steam-tubes.

When, in consequence of the fire under the boiler being extinguished or being much diminished, a vacuum begins to be formed in the steam-tubes, the external air, pressing against the surface of the small quantity of water remaining in the lower part of the cylinder *b*, forces it through the semi-circular tube *d* into the cylinder *a*, and following it into that cylinder opens for itself a passage into the steam-tubes, and prevents their being crushed by the pressure of the atmosphere, on the condensation of the steam.

When the fire is gone out, and the whole apparatus

becomes cold, the steam-tubes will be entirely filled with air.

When, on lighting the fire again, fresh steam is generated, as this steam enters the large steam-tubes in the *highest* or *most elevated* part of them, and as steam is specifically lighter than atmospheric air, the steam remains above the air which still occupies the steam-tubes, and accumulating there presses this air downwards, and by degrees forces it out of the apparatus through the same passage by which it entered; the water in the semi-circular tube supplying the place of a valve, or rather of two valves, in these different operations.

# OBSERVATIONS

CONCERNING THE

## SALUBRITY OF WARM ROOMS IN COLD WEATHER.

# OF THE SALUBRITY OF WARM ROOMS.

———

IT is a question often discussed in this country, whether living in a warm room in winter be or be not detrimental to health?

There is no doubt whatever of the necessity of *pure* air for the support of life and health; but I really do believe that erroneous opinions are entertained by many people in this island respecting the effects of that equal and at the same time moderate heat which can only be obtained in rooms where strong currents of air up the chimney are not permitted. Those who have been used to living in large apartments, in which the large fires that are kept up, instead of making the rooms equally warm, do little more than increase the violence of those streams of cold air which come whistling in through every crevice of the doors and windows,—when such persons come into a room in which an equal and genial warmth prevails in every part, struck with the novelty of the sensation that this general warmth produces, they are very apt to fancy that the air is *close*, and consequently that it must be unwholesome, and are uneasy until a door or a window be opened in order that they may get what they call *fresh* air.

But they do not seem to make a proper distinction between *fresh* air and *pure* air. When they call for *fresh* air, they doubtless mean *purer* air. They certainly get *colder* air, but I much doubt whether they often get air that is more wholesome to breathe; and it is most certain that the chilling streams and eddies that are occasioned in the room by the fresh air so introduced are extremely dangerous, and often are the cause of the most fatal disorders.

It is universally allowed to be very dangerous to be exposed in a stream of cold air, especially when standing or sitting still ; but how much must the danger be increased if one side of the body be heated by the powerful rays from a large fire, while the other is chilled by these cold blasts ? And there is this singular circumstance attending these chills, that they frequently produce their mischievous effects without our being sensible of them: for, as the mind is incapable of attending to more than one sensation at one and the same time, if the *intensity* of the sensation produced by the heat on the one side of the body be superior to that of the cold on the other, we shall remain perfectly insensible of the cold, however severe it may really be ; and if we are induced by the disagreeableness of what we do feel to turn about, or change our position or situation, this movement will be occasioned not by the cold, which we do not feel, but by the heat, which being superior in its effect upon us engages all our attention. And hence we may account for those severe colds or catarrhs which are so frequently gotten in hot rooms in this country by persons who are not conscious at the time of being exposed to any cold, but, on the contrary, suffer great and continual inconvenience from the heat.

I have said that these colds are gotten in *hot rooms*, but it would have been more accurate to have said *in rooms where there is a great fire,* or where there is a *great heat,* occasioned by a great number of burning candles, or by a great number of persons crowded together; for it is very seldom indeed that a room is much heated in this country, and their being cold is the principal cause which renders partial heats that occasionally exist in them so very injurious to health.

The air of the room that comes into contact with the cold walls, and with the enormous windows which, in open defiance of every principle of good taste, have lately come into fashion, is suddenly cooled; and being condensed, and made specifically heavier than it was before, in consequence of this loss of heat it descends and forms cold streams, that are so much the more rapid and more dangerous as the partial heats in the room are more intense. Consequently, they are the more dangerous, as they are less liable to be observed or felt.

If to these cold currents which are generated in the room, we add those which come into it from without to supply the enormous quantity of air that is continually going off by the chimney, when there is a great quantity of coals burning in an open grate, we shall not be surprised that those who venture to go in such rooms without being well wrapped up in furs, or other warm clothing, should be liable to take colds.

I never see a delicate young lady dressed in thin muslins or gauzes, in the midst of winter, expose herself in such a perilous situation, without shuddering for the consequences. But how many young persons of both sexes do we find of delicate habits, and particularly among the higher ranks of society? And

what vast numbers are carried off annually by consumptions?

It is well known that this dreadful disorder is almost
always brought on by colds, and that the cold of
winter is commonly fatal to consumptive people; but
why should the inhabitants of this island be so peculiarly subject to these colds? Is it not highly probable
that it is because they do not take proper care to prevent them? For my part, I declare, in the most serious manner, that I have not the smallest doubt that
this is really the case.

Much has been said of the supposed danger of
keeping rooms warm in winter, on account of the
necessity most people are under of sometimes going
into the cold air. But how many proofs are there that
these sudden transitions from heat to cold, or from cold
to heat, are not attended with danger, if care be taken
to be properly clothed, and if the heats and colds are
not partial?

How very hot do the Swedes and the Russians keep
their houses during the long and severe frosts that prevail in winter in those countries! And yet no people
are more strong and healthy than they are, nor are
there any less liable to catarrhs and consumptions.

It is the very warm rooms in which this hardy race
of men spend much of their time in winter (which, by
promoting a free circulation of their blood, gives them
health and strength) that enables them to support
without injury exposure *for short periods* to the most
intense cold.

In Germany the rooms of people of rank and fashion
are commonly kept in winter at the temperature of
about 64° or 65° of Fahrenheit's thermometer (the

dwellings of the peasants are kept much hotter); but though the ladies in that country are from their infancy brought up with the greatest care, and are as little exposed to hardships as the women of condition in this or in any other country, they find no inconvenience in going out of these warm rooms into the cold air. They even frequent the plays and the operas, and go on sleighing parties, during the severest frosts, and spend one whole month in the depth of winter (in the season of the carnival) in one continued round of balls and masquerades. And, what may perhaps appear to many still more incredible, they seldom fail, whatever the severity of the weather may be, to spend half an hour every morning in a cold church.

But if in Germany, where the winters are incomparably more severe than they are in this country, persons tenderly brought up, and of delicate habits, find no inconvenience whatever in living in warm rooms, and in going from them into the cold air, why should warm rooms be unwholesome in this country?

There cannot surely be any thing injurious to health in the genial warmth of 60° or 65°; and, if *pure air* for respiration is what is wanted, the great height of our rooms in England secures us against all danger from that quarter.

The prejudice in this country against living in warm rooms in winter has arisen from a very natural cause; and though the prejudice is general, and very deeply rooted, as its cause is known to me, I really have hopes that I shall be able to combat it with some success. I am perfectly sure that justice will be done to the purity of my intentions in engaging in this arduous undertaking, and *that* I look upon as a circumstance

of no small importance, especially when I consider
that it can hardly escape the observation of my reader
that few persons can be better qualified by their own
experience to give an opinion on any subject than I
happen to be to give mine on that under consideration.

I went to Germany many years ago, with as strong
a prejudice against warm rooms as anybody can have;
but, after having spent twelve winters in that country, I
have learned to know that warm rooms are very comfortable in cold weather, and that they certainly tend
to the preservation of health.

Having occupied a very large house, in which there
are several apartments that are furnished with open
chimney fire-places, I have had an excellent opportunity of making experiments of the comparative advantages and disadvantages of warming rooms with them
and with stoves; and my opinions on these subjects
have not been hastily formed, but have been the result
of much patient investigation.  They have been the
result of conviction.

Were there any thing *new* in what I recommend, I
might be suspected of being influenced by a desire
to enhance the merit of my own discoveries or inventions; but, as there is not, this suspicion cannot exist,
and I may fairly expect to be heard with that impartial-
ity which the purity of my intentions gives me a right
to expect.

It may perhaps be asked by some, what right I
have to meddle at all in a business that does not concern me personally?  Why not let the people of this
country go on quietly in their own way, without tiring
them with proposals for introducing changes in their

customs and manner of living, to which they evidently have a decided aversion ?

To such questions and observations as these I should make no reply, but should still feel anxious to promote by every means in my power all such improvements as tend to increase the comforts and innocent enjoyments of life, from whatever quarter they might come.

If it be wisdom to choose what is good, it must be folly to refuse what is advantageous to us; and, if liberality be an ornament to a respectable character, it is weakness to be ashamed of adopting the useful inventions of our neighbours.

I am not without hopes that at some future period houses in England will become as celebrated for warmth and comfort as they are now for neatness, and for the richness and elegance of their furniture.

However habit may have reconciled us to it, or rendered us insensible to its effects, cold is undoubtedly a very great physical evil. It may be, and no doubt is, productive of good in some way or other, but that is not a sufficient reason why we should not endeavour to guard ourselves against its painful and disagreeable effects. Their being painful is a proof of their being hurtful, and it is moreover a kind intimation to us of the presence of an enemy to be avoided.

We may no doubt by habit inure ourselves to cold in such a manner as to render our bodies in some degree insensible to it; but does it necessarily follow that by these means its pernicious effects on us are prevented, or even diminished ? I see no reason for supposing this to be the case.

If inuring to cold were a sufficient preservative against its bad effects, this method, which certainly would be the most economical, would, we have reason to think, have been adopted by Providence in respect to brute animals; but beasts and birds, which pass the winter in cold climates, are all furnished with warm winter garments.

What provident Nature furnishes to brute animals, man is left to provide for himself, or to supply the want of it by his ingenuity.

If living in cold rooms really tended to give strength and vigour to the constitution, and to enable men to support without injury the piercing cold of winter, we might expect that the dwellings of the inhabitants of the polar regions would be kept at a very low temperature; but this is so far from being the case in fact that we always find the hottest rooms in the coldest climates.

If the transition from a hot room to the cold air were so dangerous as it is represented, how does it happen that Swedes and Laplanders, who live in rooms that are kept excessively hot, do not take cold when they expose themselves to the intense cold of their winters?

Swedes and Russians who pass the winter in England never fail to complain of the uncomfortable coldness of our houses, and seldom escape catarrhs and other disorders occasioned by cold. And the sickness and mortality which prevailed among the Russian soldiers and sailors, who wintered in this country in the years 1798 and 1799, were generally, and no doubt justly, ascribed to their being unable to support the cold to which they were exposed in our barracks and in our hospitals, — a degree of cold to which they never had been accustomed *within doors*, and which to them appeared to be quite insupportable.

These are strong facts, and the evidence they afford in the case under consideration is pointed, and appears to me to be incontrovertible. There are many other similar facts that might be adduced in support of the position we are endeavouring to establish.

It has often been objected to warm rooms, that the air in them is always confined, and consequently unwholesome ; but no argument more perfectly groundless and nugatory was ever adduced in support of a bad cause.

When in cold weather a room is kept warm, the air in it, so far from being confined, is continually changing. Being specifically lighter, in consequence of its being warm, than the air without, it is impossible to open and shut a door without vast quantities of it being forced out of the room by the colder air from without, which rushes in ; and if at any time it be required to ventilate the room in so complete a manner that not a particle of the air in it shall remain in it, this may be done in less time than one minute, merely by letting down the top of one of the windows, and at the same time opening a door which will admit the external colder and heavier air. And it must not be imagined that the room will be much cooled in consequence of this complete ventilation. So far from it, a person returning into it, three or four minutes after it had been ventilated and the air in it totally changed, would not find its temperature sensibly altered.

The *walls of the room* would still be nearly as warm as before, and the radiant heat from those walls, passing through the transparent air of the room without any sensible diminution of their calorific powers, would produce the same sensation of warmth as they did before.   And even the cold air admitted into the room would in a few minutes become really warm.   And as the specific gravity of air is so very small, compared with that of the dense solid materials of which the walls, floor, and ceiling of the room are constructed, the warming of this air will not sensibly cool the room.

Hence we see how easy it is to ventilate warm rooms in cold weather, and also how impossible it would be to live in such a room without the air in it being perpetually changed and replaced with fresh and pure air from without.

It is those who inhabit cold rooms who are exposed to the danger of breathing confined air, for it would be in vain to open the doors and windows of such an apartment: if the air in it is as cold, and consequently just as heavy, as that without, there is no physical reason why it should move out of its place.   Part of it may, indeed, be blown out by a wind, or without opening the door and windows: a *part of it* may be forced up the chimney, if there be a fire burning in it ; but this kind of ventilation is not only dangerous in a very high degree to the health of those who are in the room, but it is also partial and very incomplete.   As the currents of cold air which supply the draught of an open-chimney fire are confined to the bottom of the room, below the level of the mantel of the fire-place, the same air may remain for weeks in the upper parts of the room, and perhaps for a much longer time in some remote corner, far from the fire.

I think enough has now been said to prove to the satisfaction of every reasonable person, who is disposed to listen and willing to be convinced, that the air in rooms properly and equally warmed in cold weather cannot be confined and contaminated; and that inhabiting warm rooms in winter, so far from rendering persons weak and unable to bear the cold on going abroad, is the best preservative against the bad effects of occasional exposure to cold.

If there are any persons who like cold rooms and partial chilling streams of cold air, and prefer them to the genial warmth of a mild and equal temperature, that choice must be considered as a matter of *taste*, about which there is no disputing.

There is a simple experiment, easily made and nowise dangerous, which shows in a sensible and convincing manner that warmth prepares the body to bear occasional cold without pain and without injury. Let a person in health, rising from a warm bed after a good night's rest in cold weather, put on a dry, warm shirt, and dressing himself merely in his drawers, stockings, and slippers, let him go into a room in which there is no fire, and walk leisurely about the room for half an hour, or let him sit down and write or read during that time, he will find himself able to support this trial without the smallest inconvenience. The cold to which he exposes himself will hardly be felt, and no bad consequences to his health will result from the experiment. Let him now repeat this experiment under different circumstances. In the evening of a chilly day, and when he is shivering with cold, let him undress himself to his shirt, and see how long he will be able to support exposure to the air in a cold room in that light dress.

There is another remarkable fact with which I was made acquainted by an eminent physician of London

(Dr. Blane), which can hardly be accounted for but on a supposition that heat prepares and enables the body to support cold. Those persons who, after having remained several years in the hot climates of India, return to reside in this country, do not feel near so much inconvenience from the cold of our climate the first year after their return as they do the second. If they would be persuaded to live in warm rooms *when they are within doors*, and make a free use of the warm bath, they *never* would feel any inconvenience from it, and they might with safety take much more exercise in the open air than they now do.

Occasional exposure to cold when the body is prepared to support it, so far from being dangerous or injurious to health, is salubrious in a high degree.

It is in order that people may be enabled to go abroad frequently, and enjoy the fine, bracing cold of winter, that I am so anxious that they should inhabit warm, comfortable rooms when they are within doors. But if, during the time when they are sitting still without exercise, the circulation of the blood is gradually and insensibly diminished by the cold which surrounds them, and above all by the cold currents of air in which they are exposed, it is not possible that they should be able to support an additional degree of cold without sinking under it.

They are like water which by long exposure to moderate cold in a state of rest has been slowly cooled down below the freezing-point : the smallest additional cold or the small agitation changes it to ice in an instant; but water at a higher temperature and full of latent heat will support the same degree of severe frost for a considerable time without appearing to be at all affected by it. The more attentively this comparison is considered, the more just will it be found, and the more conclusive will be the inferences that are derived from it.

If man has been less kindly used than brute animals, by being sent naked into the world without a garment to cover and defend from the inclemency of the seasons, the power which has been given him over FIRE has made the most ample amends for that natural deficiency; and it would be wise in us to derive all possible advantages from the exercise of the high prerogative we enjoy.

# OBSERVATIONS

CONCERNING THE

## SALUBRITY OF WARM BATHING

AND

## THE PRINCIPLES ON WHICH WARM BATHS SHOULD BE CONSTRUCTED.

# OF THE SALUBRITY OF WARM BATHING.

------

HAD I any hopes of being able by any thing I could say to prevail on the inhabitants of this island to adopt more generally a practice which so many nations have considered as a most rational luxury, and which, no doubt, is as conducive to health as it is essential to personal cleanliness, I should think my time well employed, were I to write a volume in recommendation of warm bathing; but I am sensible that, after all that has already been said on that subject by ancient and modern writers, — by historians and by medical men, — what I could add would be of little avail. The subject is, however, so intimately connected with that treated in the preceding Essay, that I may, perhaps, without any impropriety, take the liberty to make a few observations concerning it.

If a perfectly free circulation of the blood, brought on and kept up for a certain time without any violent muscular exertion, and consequently without any expense of strength, be conducive to health, in, that case warm bathing must be wholesome; and, so far from weakening the constitution, must tend very powerfully to strengthen it.

Among those nations where warm bathing has been most generally practised, and where the effects of it have, of course, been best known, no doubts have ever been entertained of its being very beneficial to health ; and nobody can doubt of its being pleasant and agreeable in a high degree.

Had warm bathing never prevailed but in certain climates, doubts might be entertained of its *general* usefulness ; but so many nations, remote from each other, and inhabiting countries extremely different, not only in respect to climate, but also in respect to situation and produce, and where manners and customs have been extremely different in all other respects, have practised it, that we may safely venture to pronounce warm bathing to be useful to man.

It was by accident I was led, about two years ago, to consider this subject with that attention which it appears to me to deserve ; and I then made an experiment on myself, the result of which I really think very interesting, and of sufficient importance to deserve being made known to the public.

The waters of Harrowgate, in Yorkshire, having been recommended to me by my physician, I went there in the month of July, 1800, and remained there two months. I began with drinking the waters at the well every morning, and with bathing in them, warmed to about ninety-six degrees of Fahrenheit's thermometer, every third day at my lodgings.

At first I went into the bath at about ten o'clock in the evening, and remained in it from ten to fifteen minutes, and immediately on coming out of it went to bed, my bed having been well warmed, with a view to preventing my *taking cold.*

Having pursued this method some time, and finding myself frequently feverish and restless after bathing, I accidentally in conversation mentioned the circumstance to an intelligent gentleman who happened to lodge in the house, and who had long been in a habit of visiting Harrowgate every year. He advised me to change my hour of bathing, and to stay longer in the bath, and, above all, to avoid going into a warmed bed on coming out of it. I followed his advice, and shall have reason all my life to thank him for it.

I now went into the bath regularly every third day, about two hours before dinner, and stayed in it half an hour, and on coming out of it, instead of going into a warmed bed, I merely had myself wiped perfectly dry with warmed cloths in a warmed room adjoining to the bath; and dressing myself in a bed-gown, which was moderately warm, I retired to my room, where I remained till dinner-time, amusing myself with walking about the room, and with reading or writing, till it was time to dress for dinner.

The good effects produced by this change of method were too striking not to be remarked and remembered. I was no longer troubled with any of those feverish heats after bathing which I experienced before; and so far from feeling *chilly*, or being particularly sensible to cold on coming out of the bath, I always found myself less sensible to cold after bathing than before. I even observed, repeatedly and invariably, that the glow of health and pleasing flow of spirits, which resulted from the full and free circulation of the blood which bathing had brought on, continued for many hours, and never was followed by any thing like that distressing languor which always succeeds to an artificial increase of circulation and momentary flow of spirits which are produced by stimulating medicines.

I regularly found that I had a better appetite for my

dinner on those days when I bathed than on those
when I did not bathe, and also that I had a better
digestion and better spirits, and was stronger to endure
fatigue, and less sensible to cold in the afternoon and
evening.

As these favourable results appeared to be quite regular and constant, I was induced to proceed to a more
decisive experiment. I now began to bathe every
*second day*, and, finding that all the advantageous
effects which I had before experienced from warm
bathing still continued, I was encouraged to go one
step further, and I now began to bathe *every day*.

This experiment was thought to be very hazardous
by many persons at Harrowgate, and even by the
physician, who did not much approve of my proceedings; but as no inconvenience of any kind appeared to
result from it, and as I found myself growing stronger
every day, and gaining fresh health, activity, and spirits,
I continued the practice, and actually bathed *every day*
at two o'clock in the afternoon for half an hour in a
bath at the temperature of 96° and 97° of Fahrenheit's scale, during *thirty-five days*.

The salutary effects of this experiment were perfectly evident to all those who were present and saw
the progress of it, and the advantages I received from
it have been permanent. The good state of health
which I have since enjoyed I ascribe to it entirely.
But it is not merely on account of the advantages
which I happened to derive from warm bathing which
renders me so warm an advocate for the practice.
Exclusive of the wholesomeness of the warm bath, the
luxury of bathing is so great, and the tranquil state of
mind and body which follows it is so exquisitely de-

lightful, that I think it quite impossible to recommend it too strongly, if we consider it merely as a rational and elegant refinement.

I am persuaded, however, that we are very far in this country from understanding the best method of fitting up warm baths, and of using them in the most comfortable and advantageous manner. It appears to me to be quite evident that it is not the water, but the *warmth*, to which most, if not all, the good effects experienced from warm bathing ought to be ascribed.

Among those nations where warm bathing has been most generally practised, water has seldom been employed, except occasionally, and merely for washing and cleaning the skin; and though washing in warm water is pleasant, and is, no doubt, very wholesome, yet remaining with the whole body, except the head, plunged and immersed in that liquid for so great a length of time as is necessary, in order that a warm bath may produce its proper salutary effects, is not very agreeable, nor is it probably either necessary or salutary.

The manner in which a warm bath operates in producing the pleasant and salutary effects which are found to be derived from it appears to me to be so evident as to admit of no doubt or difference of opinion on that subject.

The genial warmth which is applied to the skin, in the place of the cold air of the atmosphere by which we are commonly surrounded, expands all those very small vessels where the extremities of the arteries and veins unite, and, by gently stimulating the whole frame, produces a free and full circulation, which, if continued for a certain time, removes all obstructions in the vascular system, and puts all the organs into that state of regular, free, and full motion which is essential to health, and also to that delightful repose, accompanied by a consciousness of the power of exertion, which constitutes the highest animal enjoyment of which we are capable.

If this statement be accurate, it cannot be difficult to explain, in a manner perfectly satisfactory, why a warm bath is often found to produce effects when first used, and especially by those who stay in the bath for too short a time, which are very different from those which it ought to produce, and which it cannot fail to produce when properly managed. We shall likewise be enabled to account for the feverish symptoms which result from going out of a warm bath into a warmed bed.

The beginning of that strong circulation which is occasioned on first going into a warm bath is an effort of Nature to remove obstructions; and if time be not given to her to complete her work, and if she be checked in the midst of it, the consequences must necessarily be very different from those which would result from a more scientific and prudent management. Hence we see how necessary it is to remain in a warm bath a sufficient time; and, above all, how essential it is that the bath should be *really warm*, and not tepid, or what has been called *temperate*.

When we consider the rapidity with which water carries off heat from any body hotter than it which is immersed in it, we shall find reason for astonishment that any person, even the strongest man in a state of the highest health, is able to support the loss of heat which must necessarily result from lying for half an

hour quite motionless in a tub of water at the temperature of 55 or 60 degrees; and yet, if I am rightly
informed, baths at that temperature have sometimes
been ordered by physicians, and even for persons of
delicate constitutions.

Because we are able to support that degree of cold
without injury *in air*, that is very far indeed from
being a good reason for concluding that *water* at
that temperature would not be hurtful; for water is
800 times more dense than air, and consequently when
it is cold must deprive our bodies of heat when we are
immersed in it, with infinitely greater rapidity than air
at the same temperature can do.

Having reason to think that physicians in general
are not sufficiently aware of the very great difference
there is in the powers of these two fluids to *carry off*
heat when they are both at the same temperature,
and having myself been a witness more than once to
very alarming consequences which have resulted from
the use of what was called a *tepid bath*, I cannot resist
the inclination I feel to avail myself of this opportunity
of calling the attention of medical men to a circumstance which is most undoubtedly of very serious
importance.

When we go into a bath at the temperature of about
96 degrees (which is blood heat), though the water at
first may seem warm to us, and even hot, yet it is not
capable of communicating much heat to us: for our
bodies being at the same temperature, except it be
perhaps at the very surface of the skin (where the
nerves of feeling are most plentifully distributed), there
is no reason why heat should pass out of the water
into us; but if the water be only a few degrees below

the temperature of the blood, though it may feel warm when we first go into the bath, yet that sensation will soon be followed by one of a very different nature, and the water will carry off heat very rapidly from the surface of the body.

A rapid cooling of the body, by carrying off by a mechanical process the heat generated in the body by the action of the vital powers, may or may not be advisable in certain cases. That is a question of nice discrimination, and one upon which I am perfectly sensible that I am not qualified to decide; but I may be allowed to point out physical consequences not very obvious, and consequently not likely to be subjects of meditation and investigation, which ought certainly to be rightly understood.

There is one observation more respecting tepid and temperate baths which appears to me to be deserving of particular attention; and that is the state of *inaction* in which a person commonly remains in such a bath, and the probable consequences of inaction under such circumstances. Swimming is universally allowed to be a wholesome exercise, and there are few instances, I believe, of harm arising from it, even when the water has been at a much lower temperature than that of the blood; but I am far from being of opinion that remaining in the water without any muscular exertion would be found to be equally conducive to health.

Cold baths are perfectly different from hot baths and tepid baths, and the intention of the physician in ordering them is also different. I am not prepared to explain the physical effects produced by a momentary plunge into cold water, and much less to give an opinion respecting the salubrity of the practice of cold bathing, or of its usefulness as a remedy for certain diseases.

But to return from these speculations to more interesting details, — to the results of actual experiments. During the thirty-five days that I continued to make daily use of a warm bath, I made a number of experiments on myself, in order fully to satisfy my own mind on several important points respecting which I still had doubts remaining. Some of those experiments were certainly too hazardous to be reconciled to sober good sense, and to that prudent attention to the preservation of health which every wise man would be ashamed of neglecting. But though I may be blamable for my temerity, and may even expose myself to ridicule by making a discovery of my rashness, yet I am so deeply impressed with the importance of the results of some of my experiments that I cannot refrain from laying them before the public.

Having long entertained an opinion that the most effectual means that can be used to prepare the body to support, without inconvenience and without injury, those occasional exposures to cold to which every person is liable who inhabits a cold country, is, by a proper application of warmth and without the fatigue of violent muscular exertion, to bring on, and keep up for a certain time, at certain intervals, such a full, strong, and free circulation and perspiration as shall effectually remove from time to time all those gradual contractions and obstructions which chilling cold naturally produces, and give a new impulse to those actions in which life, health, and strength consist; I imagined that, if this opinion was well founded, the use of the warm bath, instead of rendering my habit more delicate, and making me more liable to take cold on exposing myself in the cold air, I should certainly find myself strengthened by it, and my constitution rendered more robust.

The first direct proofs I had that this advantageous change had actually taken place in me were accidental, and it was probably that discovery which induced and encouraged me to expose myself voluntarily to more severe trials.

I had, from the time of my first arrival at Harrowgate, been in a habit of retiring to my room towards evening every day, where I commonly spent an hour or more in reading or writing; and, as I never had any fire in my room, I frequently felt myself quite chilled by the cold of the evening. At this time I bathed only once in three days; but, after I had begun to go into the bath before dinner, I soon found that I was much less sensible to the cold of the evening on those days when I bathed, than on those when I did not bathe.

It was the discovery of this interesting fact which contributed much, and perhaps more than any thing else, to induce me to take the resolution (which was considered as very violent and unadvised) of going into the bath every second day, and afterwards every day.

After I had continued to bathe every day for some time, I no longer felt the smallest inconvenience from the cold of the evening, though I frequently sat in my room with the windows open when the weather was very cold and chilly, till it was so dark that I could neither see to read nor to write; and when I joined the company below I felt myself in high spirits, and never wanted an excellent appetite to my supper. My sleep was undisturbed and refreshing, and every thing indicated the return of perfect health.

All these favourable appearances having continued for some time, and finding my strength to increase daily, I became more venturous, and frequently went out after it was dark, when the evening was cold and raw, and walked alone more than half an hour on the bleak, dreary common which lies before the house where I lodged (the Ganby Inn), to see if my constitution was really so much changed as to enable me to support that trial without taking cold.

I even returned on foot from the play-house, across the common, several times in the evening, lightly dressed, when a cold wind blew over the common, and after I had suffered much from heat in the theatre; but in none of these severe trials did I receive the smallest injury. I never took cold, nor did I experience any feverish heats or restlessness on going to bed after them. I call them *severe* trials, and as such they will doubtless be considered, when it is recollected that, when I arrived at Harrowgate, I was far from being in a good state of health (having never recovered from the dangerous illness I had brought on myself six or seven years before in Bavaria, by excessive application to public business), and when it is remembered that at the time when I was exposing myself in this manner to the danger of taking cold I was using the warm bath every day.

But I am firmly persuaded that it was to the *warm bath* that I was indebted for my escape; and it is that persuasion which has induced me to publish this account of my experiment.

I am very far, indeed, from wishing that my example should be followed in all points. All the unadvised and imprudent details of the experiment may, and ought to be, omitted. It would, indeed, be more than imprudent — it would be foolish — to repeat them. But I do really believe that all those who will be persuaded to adopt the practice of warm bathing, in health and in sickness, will find the greatest and most permanent advantages from it.

Were the general and constant use of the warm bath by persons in health a new thing, I should have many scruples in recommending it to the public, whatever my private opinion of its salubrity might be. But so many nations have practised it for ages, and there are so many who now practise it, and, what is very remarkable, one (the Russian) which inhabits the coldest parts of the globe, that there cannot possibly be the smallest reason to doubt of its beneficial effects.

With regard to the *pleasant* effects that result from the use of the warm bath, there never has been any difference of opinion. But still I am quite certain that the true luxury of warm bathing is not understood in this country; and, till the construction of our baths is totally changed, and a different manner of using them adopted, we never can enjoy a warm bath as it ought to be enjoyed.

As we must allow that in most cases, and particularly in a matter of this kind, it is much more wise and prudent to adopt those arrangements and improvements which have been the result of the experience of ages than to sit down and attempt to invent any thing new, I think we cannot do better than to rebuild some of the baths which were left us by the Romans. *They* most certainly understood warm bathing as well as any nation ever did; and, if there be any thing in our climate which renders any deviations necessary from the manner commonly practised in constructing baths in warmer countries, there is no doubt but those luxurious foreigners, who had possession of this island for so many years, must have found them out. The plans they have left us may therefore be adopted with safety as models for our imitation.

I am far from wishing to see the baths of Diocletian and Caracalla rise up in all their splendour in the neighbourhood of London; for I am well aware that the magnificent and ostentatious exhibitions of a nation of conquerors and slaves would but ill accord with the manners of a free, enlightened, and industrious people; but still I cannot help wishing that the inhabitants of this island, and all mankind, might enjoy all the innocent luxuries and comforts that are within their reach.

I am even jealous of the poor Russian peasant; and when I see him enjoying the highest degree of delight and satisfaction in the rude cave which he calls a warm bath, without wishing to diminish his pleasure, I greatly lament that so useful and so delightful an enjoyment should be totally unknown to so great a portion of the human species.

Who knows but that the poor Russian, in the midst of his snows, with his warm room and warm bath, may not, on the whole, enjoy quite as much happiness as the inhabitant of any other country? And, if this be really the case, what an addition would it be to the enjoyments of the inhabitants of other more favoured countries to add the warm room and warm bath of the

Russian to all their local advantages! When I meditate profoundly on these subjects, it is quite impossible
for me not to feel my bosom warmed with the most
enthusiastic zeal for the diffusion of that knowledge
which contributes to the comforts and enjoyments
of life.

There is nothing more interesting than the results
of the ingenuity of man in the infancy of society, before the light of science has extended his views and increased the number of the objects of his pursuit. Ever
intent upon a few simple mechanical contrivances, the
usefulness of which he continually experiences, all his
thoughts remain concentrated on them, and all his ingenuity and address are employed in rendering them
perfect, and using them with agility and effect. When
we examine the implements which savage nations have
contrived to provide for themselves, almost without
tools, we shall see one of the most striking proofs to
be found of the effects of persevering industry and
long experience.

No person of any feeling can contemplate the
canoes, snow-shoes, and hunting and fishing tackle of
the North American savages, without experiencing
emotions which it would be very difficult to describe;
and the ingenuity displayed by the Russian peasant in
the construction and management of his warm bath is
not less striking.

Without any knowledge of the principles of pneu-
matics, hydrostatics, and chemistry, he has proceeded
in the same manner precisely as he would have done
had he understood all those sciences; and, without
money or the means of purchasing any thing of value,
he has contrived, with the rude materials of no value

which he finds lying about him, to construct an edifice in which he enjoys, in the most complete manner possible, all the delightful sensations which result from one of the most rational pleasures of the most refined and luxurious nations. And if security in the possession of an advantage adds value to it, how much greater is the security of the Russian peasant in the enjoyment of his luxuries than the rich and effeminate in the possession of theirs! Nothing is more calculated to fill us with wonder and admiration than to see how the different situations of man on this globe have been equalized by compensations.

The warm baths of the Russian peasants have so often been described, that I dare not take up the reader's time unnecessarily by giving a particular account of them. They are, as is well known, what are called vapour baths; and, as those who build them are much too poor to afford the expense either of boilers or bathing-tubs, they are heated in a manner which is equally ingenious and economical. A parcel of stones are heated upon a wood fire made on the ground, and, when these stones are hot, water or snow is thrown on them, and the steam which is produced rises up and occupies the inside of the arched roof of the cave which constitutes the bath.

Those who enjoy the bath place themselves, extended at full length, on a bed composed of the small twigs and leaves of trees, on hurdles in the form of shelves, placed round the cave under its vaulted roof, and above the level of the top of the door-way.

From this short description, it is evident that the air occupying the top of the cave, and which is heated by the steam, being rendered specifically lighter than the cold air without by the heat it has acquired, will remain in its place, even though the entrance into the cave should not be provided with a door. A few branches of trees, placed against the door-way, would break the force of the wind, if any were stirring; and the bath would remain as warm as should be required for any length of time, even in the most severe frost of a Russian winter, with the expense of a very small quantity of fuel.

Were I asked to give a plan for a warm bath by a friend who had full confidence in my abilities to execute such an undertaking with intelligence, I should adopt, with little deviation, all the principles of the Russian baths.

The bath-room should be built of bricks, and should be covered above by a Gothic or pointed dome; and the entrance into it should not be through the side walls, but through the pavement, by a flight of steps from below. The walls should be double, the inner wall being made as thin as possible; and the room should be lighted by three or four very small double windows, of single panes of glass, situated just below the spring of the dome, which might be at the height of seven or eight feet above the pavement.

As the (double) walls of the building would be of some considerable thickness, and as the windows ought to be small and double, it would be very easy to construct them in such a manner that a person from without should not be able to see any person in the bath, even though they were to get a ladder and attempt to look in at the window. One of the windows should be made to open, in order to ventilate the bath.

The inside of the walls and dome of the bath-room

should be plastered, and afterward well painted in oil, or, what would have a neater and more elegant appearance, they might be lined with Dutch tile.

The pavement might be made of any kind of flat stones, or of bricks or tiles; or it might be constructed of stucco, well painted in oil, and it might be covered with matting.

If ornament were required, I would place a figure of Vesta, holding an Argand's lamp, on a pedestal, on one side of the room. This pedestal, which should be large in proportion to the figure, should be made of sheet-copper, and painted of a bronze colour on the outside. The cavity within it should be accurately closed on every side, in order that it might occasionally be filled with steam from a boiler situated without, and used as a stove for warming the room.

The important object had in view in making the entrance into this bath from below (the preservation of the warm air in the room) might be attained equally well with the door placed on one side of the room, provided the door were made to open immediately into a narrow, descending, vaulted gallery, furnished with a good door at the lower end of it.

The top of the door at the lower end of this gallery should be two or three feet below the level of the bottom of the door at the top of it, which opens into the bath.

By setting both these doors open, and at the same time opening one of the windows of the bath, all the warm air in it, below the level of the window, will be forced out in a very few moments, and the room will be completely ventilated.

If the entrance be made through the side of the

room, in the manner just described, this will render the form of the room more simple and more elegant than if the passage into it were from below, through the pavement.

If the pavement of the bath be on a level, or nearly on a level, with the surface of the ground, the entrance into it must, nevertheless, come from a lower place. If the door leading into the bath be situated at one side of the room, the vaulted gallery with which it communicates must descend below the level of the surface of the ground, and a passage must be opened from without, in order to arrive at the door which must close this gallery at its lower extremity.

A steam-boiler should be placed under the bath, in a vaulted room, and the smoke from the closed fireplace of the boiler should be made to circulate in flues under the pavement of the bath, near the walls of the room, in which part the pavement should not be covered with matting.

A bathing-tub should stand on one side of the room, and opposite to it should be placed a bamboo or caned sofa, covered first with a soft, thick blanket, and then with a clean sheet thrown over it.

The bathing-tub, which might be of the usual dimensions, should be placed on a platform of wood, covered with sheet-lead about seven or eight feet square, and raised six or seven inches above the pavement. This platform should be flat and nearly horizontal, with a border all round it about two or three inches high, and a leaden pipe at the lowest part of it to carry off the water that happens to fall on it.

The lead should be covered by thin boards, or by a loose piece of matting; and a caned chair or a stool should be placed on the platform by the side of the bathing-tub. A pipe should be prepared for admitting cold water into the bathing-tub from a reservoir situated without the bath; and another, for bringing steam into it to heat it from the steam-boiler. There should likewise be a waste-pipe for carrying off the water when the bathing-tub is emptied.

The bathing-tub should not be set down immediately upon the lead which covers the platform on which the tub is placed, but should be raised eight or ten inches above it, in order that the air may pass freely under the bottom of the tub, and that there may be room to come at the lead to wash it and clean it in every part.

A bath, constructed in the manner here described, might be kept constantly warm all the year round, at a very small expense for fuel; and in that case it would always be ready for use.

It is equally well calculated to serve as a warm air-bath, as a vapour-bath, or as a warm-water bath; and, when it is used as a water-bath, the air in the room may be made either warm or temperate at pleasure.

This last circumstance I take to be a matter of the greatest importance; for nothing surely can be more disagreeable than the sensations of a person on getting out of a tub of warm water, and standing shivering with cold till he is wiped dry and dressed; and I cannot help suspecting that such a situation is as dangerous as it is unpleasant.

I am much inclined to think that the warm *air-bath*, with occasional washing with warm water, will be found to be not only the most pleasant, but also the most wholesome, of any; and, if that should be the case, no building could answer for that purpose in this country (where the temperature of the atmosphere is always so much below that which would be wanted), unless it were constructed on principles similar to those on which the plan above described is founded.

*Hot air* may at any time be procured in any climate; but a large mass of air moderately and *equally* warm cannot be *preserved,* in a cold country, by any other means than by preventing its being cooled, and preventing its being driven away by the denser surrounding medium.

The double walls and small double windows of the bath which I have recommended will prevent the *cooling* of the air in it; and the form of the room renders it absolutely impossible for the cold air of the atmosphere either to mix with that warm air or to *force it out of its place.*

If it be required to mix steam with the air of the room to render it moist, that may be done by laying a steam tube, for that purpose, from the boiler into the room; or it may be done in a manner still more refined and luxurious, by having a small portable boiler for that purpose, heated by a spirit lamp; or a common tea-urn heated or rather kept boiling by an iron heater, or a common tea-kettle heated by a spirit lamp, might be made use of. The water might be brought in already boiling hot, and, if a quantity of cloves or other spices were mixed with it, the room would be filled with the most grateful and most salutary perfumes. By burning sweet-scented woods or aromatic gums and resins in the room, in a small chafing-dish filled with live coals, the air in the room would be perfumed with the most pleasant aromatic odours.

Those who are disposed to smile at this display of Eastern luxury would do well to reflect on the sums they expend on what *they* consider as luxuries, and then compare the real and *harmless* enjoyments derived from them with the rational and innocent pleasures here recommended. I would ask them, if a statesman or a soldier, going from the refreshing enjoyment of a bath such as I have described to the senate or to the field, would, in their opinion, be less likely to do his duty than a person whose head is filled, and whose faculties are deranged, by the fumes of wine.

*Effeminacy* is no doubt very despicable, especially in a person who aspires to the character and virtues of a man ; but I see no cause for calling any thing *effeminate* which has no tendency to diminish either the strength of the body, the dignity of sentiment, or the energy of the mind. I see no good reason for considering those grateful aromatic perfumes, which in all ages have been held in such high estimation, as a less elegant or less rational luxury than smoking tobacco or stuffing the nose with snuff.

Having given a slight sketch of a bath on a scale of magnificence and refinement which will not suit every person's circumstances, and may not accord with every person's taste, I will now give another on a less expensive and more modest plan.

Let a small building be erected 14 feet 5 inches long and 9 feet wide, measured within, and 7 feet high ; and let it be divided into equal rooms of 9 feet long and 7 feet wide each, by a partition wall of brick 4½ inches wide, or equal in thickness to the width of a brick. Let the outside walls of this little edifice be double, the two walls being each the width of brick in thickness, and the void space between them being likewise of the same thickness; viz., about $4\frac{1}{2}$ inches. In order to strengthen these double walls, they may be braced and supported one against the other, by uniting them in different parts by single bricks laid across, with their two ends fixed in the two walls.

Instead of a floor of boards, these two little rooms should be paved with 12-inch tiles or flat stones, laid in such a manner, on thin parallel walls ($4\frac{1}{2}$ inches in thickness), as to form horizontal flues under every part of the pavement.

There should be no door of communication between these rooms; but each should have its separate entrance from without, by a door opening directly into a separate narrow, descending, covered gallery. These two doors should be placed on the same side of the building; and their two separate descending galleries may be parallel to each other, and may indeed be covered by the same roof.

They may together form one gallery, divided into two narrow passages by a thin partition wall constructed with bricks.

A small porch at the bottom of the gallery should be common to both passages; but each passage should nevertheless have its separate door at its lower extremity, where it communicates with the porch.

The top of the door-way of this descending passage at its lower extremity must be at least one foot below the level of the pavement of the rooms.

This passage may be furnished with a flight of steps, or its descent may be made so easy as to render steps unnecessary.

If there should be no natural elevation of ground at hand on which this bath can conveniently be situated, a mound of earth must be raised for that purpose; otherwise, it will be necessary that the porch at the end of the gallery should be situated 7 or 8 feet below the surface of the ground, for it is indispensably necessary that the entrance into the bath should be by an *ascent*, and in a *covered gallery.*\*

The building may be covered with a thick, thatched roof, which will on some accounts be better than any other; but any other kind of roof will answer very well, provided it be tight, and that a quantity of straw or of chaff or of dry leaves be laid over the ceiling of the two small rooms, under the roof, to confine the heat. The ceiling of the rooms should be lathed and plastered, and the walls of the room should be plastered and whitewashed.

At the end of one of the rooms, opposite to the door, a bathing-tub should be placed; and in the other, a caned sofa.

The bathing-tub should be placed on a platform 7 feet square, covered with sheet-lead, and raised about nine inches above the level of the pavement. This platform should have a rim all round it, and a pipe for carrying off out of the room the water that accidentally falls on it.

The bathing-tub should be supplied with cold water from a reservoir (a common cask will answer perfectly well for that use), which should stand without the house.

---

\* If the entrance into the houses of poor cottagers were constructed on the same principles, this simple contrivance would save them more than half their expenses for fuel in cold weather.

The water should be admitted cold into the bathing-tub, and should be warmed in it by means of steam, which may come from a small steam-boiler, which should be situated without the building and near to the reservoir of cold water. A small open shed, made against one side of the building, — that side of it which is opposite to the entrance gallery, — may cover both the boiler and the reservoir. The boiler, which need not be made to contain more than six or eight gallons, should be well set in brick-work, and well covered over with bricks, to prevent the loss of heat which would result from any part of the boiler being exposed naked to the cold air of the atmosphere.

This boiler should be so fitted up by means of a ball-cork, as to feed itself regularly with water from the neighbouring reservoir.

The boiler should be furnished with a safety-valve, opening into the open air, and with a tube for conveying steam into the bathing-tub. This tube, which may be a common leaden pipe about half an inch in diameter, should be wound round with the list of coarse cloth, or with any warm covering of that sort, to confine the heat.

This steam tube should rise up perpendicularly from the boiler to the height of eight or ten inches above the level of the ceiling of the bath-room, and should then be bent towards the building, and made to enter the roof of it, and then to descend perpendicularly through the ceiling of the bath-room, and enter the bathing-tub.

Its open end should reach to within an inch of the bottom of the tub; and a little above the level of the top of the tub there should be a steam-cock, by means of which the passage of the steam through the steam tube, and into the water in the bathing-tub, may be regulated, or prevented entirely, as the occasion may require.

There may be a short branch six or eight inches long, inserted into the steam tube just described, which branch will serve for admitting steam into the room when it is designed to be used as a steam or vapour bath. This short branch must of course be furnished with its own separate steam-cock.

The smoke from the (closed) fire-place of the boiler must be made to circulate under the pavement of the two rooms of the bath, in the flues constructed for that purpose, before it is suffered to pass off into the chimney.

The chimney should stand on the outside of the building, and be made to lean against and be supported by the wall of the building. There should be a damper in this chimney.

Each of the small rooms should be furnished with a small double window; each window consisting of one large pane of glass, and being made to open by means of a hinge placed on one side of it.

These windows should be placed as near the ceiling of the room as possible, in order to facilitate the perfect and speedy ventilation of the bath. The inside windows may be placed level with the inside of the wall of the house; and the outside windows, level or flush with the outside wall. Either the inside windows or the outside windows should be made of ground or of wavy glass, in order that a person in the bath may not be exposed to being seen through the windows.

The two small rooms may be distinguished by calling one of them the *bath-room* and the other the *dressing-room.*

If it be required to heat the two rooms in a very short time, the one with vapour, and the other with dry air equally warmed and perfectly free from all disagreeable smells, this may be done by the following simple contrivance: Let a cylinder of very thin copper, about eight inches in diameter and five feet in length, be placed horizontally under the sofa in the dressing-room, and let a steam-pipe from the boiler be laid into it, with another pipe for carrying off the water resulting from the condensation of the steam in it. By admitting steam into this tube, the air in the room will soon be warmed, without any watery vapour being mixed with it; and by admitting steam into the bath-room, and, allowing it to mix with the air of that room, a vapour-bath will be formed, and in a very few minutes will be ready for use.

A small quantity of cold water may then be admitted into the bathing-tub, and, the steam being turned into it, it will soon be made warm enough to be used for washing, after the steam-bath has been used.

The passage from the bath-room into the dressing-room will be attended with no danger from cold; and it will be found very pleasant to dress and repose in a warm room, where the air is pure and not charged with vapour, after coming out of the water or out of a vapour-bath.

If there should be any apprehension that either the bath-room or the dressing-room might be too much heated by the smoke from the boiler passing continually through the flues under the pavement, a canal, furnished with a damper, leading from the closed fire-place of the boiler immediately into the chimney, might be made ; and, whenever the pavement should become too hot, by opening this canal the smoke would pass off immediately into the chimney by the shortest road, and the pavement would receive no more heat from it. I think it would in all cases be advisable to take this precaution, in constructing a bath on the principles here recommended.

But I must hasten to finish this long dissertation ; and I shall conclude it with a few passages from a modern traveller (M. SAVARY), who may be considered as being well qualified to give an opinion on the subject in question.

Speaking of the manner of using the warm bath in Egypt, he says : " The bathers here are not imprisoned, as they are in Europe, in a kind of tub where one is never at one's ease. Extended on a cloth spread out, with the head supported by a small cushion, they can stretch themselves freely in every posture, whilst they lie quite at their ease, enveloped in a cloud of odoriferous vapours, which penetrates all their pores. In this situation they repose for some time, till a gentle moisture upon the skin appears, and by degrees diffuses itself over the whole body. A servant then comes and *masses* them (as it is called, from a word in the Arabic language, which signifies *to touch in a delicate manner*). He seems to knead the flesh, but without causing the smallest pain ; and, when that operation is ended, he puts on a glove made of woollen stuff, and rubs the skin for a considerable time.

" During the whole of this time the sweat continues to be most profuse, and a considerable quantity of scaly matter and other impurities which obstructed the pores of the skin are removed, and the skin becomes quite soft, and as smooth as satin.

" When this operation is ended, the bather is conducted into a closet, in which there is a cistern supplied with hot and with cold water, which comes into it through two separate pipes, each furnished with a brass cock. Here a lather of perfumed soap is poured over him.

" After being well washed and wiped, a warm sheet is wrapped round him, and he follows the attendant, through a long winding passage, into an external and more spacious apartment. This transition from heat to cold produces no disagreeable sensations nor any bad consequences.

" In this airy apartment a bed of repose is found prepared, and fresh and dry linen is brought. A pipe is also brought, and coffee is served.

" Coming out of a hot bath, where one was surrounded by a cloud of warm vapours till the sweat gushed from every pore, and being transported into the free air of a spacious apartment, the breast dilates, and one breathes with voluptuousness. The pores of the body being perfectly cleaned and all obstructions removed, one feels, as it were, regenerated, and one experiences an universal comfort. The blood circulates with freedom, and one feels as if disengaged from an enormous weight, with a sense of suppleness and lightness which is as new as it is delightful. A lively sentiment of existence diffuses itself over the whole frame, and the soul, sympathizing in these delicate sensations, enjoys the most agreeable ideas. The imagination, wandering over the universe, which it embellishes, sees on every side the most enchanting pictures — everywhere the image of happiness!

" If the succession of our ideas be the real measure of life, the rapidity with which they then recur to the memory, and the vigour with which the mind runs over the extended chain of them, would induce a belief that in the two hours of delicious calm that succeeds the bath one has lived a number of years!"

# ON THE SPECIFIC GRAVITY,
# DIAMETER, STRENGTH, COHESION,
# ETC., OF SILK

Munich 24 June, 1786

*To Sir Joseph Banks*

Dear Sir

In the prosecution of a course of Experiments, which I began several years ago upon the force of cohesion, or *strength*, of various bodies (an account of which I intend soon to publish) in order to ascertain the volume, or solid contents of a cylinder formed of a number of single threads of raw silk, whose strength I was desirous of determining, not being able, on account of their extreme fineness, to measure the diametres of these threads, I had recourse in this as in many other like cases, to the well known method of computing the solidity of bodies from their weights and specific Gravities; But as I could not find silk in any of the tables of the bodies whose specific gravities had been determined it became necessary for me to determine its specific gravity myself; and it was in the prosecution of the various experiments which I was obliged to make for that purpose that I was led to the train of observation and experimental Enquiry which will be the subject of this Letter. And as in Philosophical Researches the history of observations leading to any general conclusion or new discovery is always interesting, and is often necessary in

433

order to our judging with greater certainty of the phenomena described, and of the consequences drawn from
them, I shall give an account in detail of these, my experiments, not only in the same order in which they were
made, but in the same form in which they were entered
and now stand registered in my diary; and accompanied
with my original observations upon them.

"London, January 20th 1781"

"In order to determine the specific gravity of silk I
made the following Experiment."

"Having procured a skain of *Raw silk* the product of
Italy, and of the best quality, I weighed it in my Hydro-
statical ballance and it was found to weigh *in air* $73\frac{3}{8}$
Grains. The same being afterwards thoroughly wet in
pump water, and being weighed *in water*, in the Glass
bucket belonging to the ballance, it was found to weigh
$17\frac{1}{4}$ Grains; but being immediately after taken out of the
bucket and weighed loose in the water it was found to
weigh no more than $15\frac{11}{16}$ grs."

If I was surprized at this apparent difference of weight
in the silk when weighed in the bucket and when weighed
loose in the water I was still more so the next day when,

"The Silk having been left 18 hours in the water was
again weighed *in water* and was found to weigh when
weighed loose in the water $14\frac{5}{8}$ grs but when weighed in
the bucket — $15\frac{5}{8}$ grs."

These extraordinary appearances awaked my curiosity,
but my time being much taken up at that period with
other avocations (being then undersecretary of state) I
had not leisure to pursue the matter further, I therefore
contented myself for the time with finishing the experiment, by drying the silk, and weighing it again *in air* to

see if its weight remained the same as before it was put into water, and with determining from the result in the best manner I was able, *the specific gravity of Silk*; the original object of the Experiment.

"The silk being taken out of the water was hung up 24 Hours in the open air, in the shade; and being afterwards more thoroughly dried before the fire, it was weighed and was found to weigh $69\frac{1}{4}$ grs."

"January 25th. 11 o'Clock A.M. The silk having now been hung up in the air in a dry room more than two days, was again dried at the fire thoroughly, after which, being immediately weighed it was found to weigh no more than $68\frac{7}{16}$ grs."

"Taking this ($68\frac{7}{16}$ grs) for its weight *in air* and taking its weight *in water* = $17\frac{1}{4}$ grains its specific Gravity turns out to that of water as $1337 = 1000$."

"The number of threads in this skain of silk being counted they were found to be 2800; and each thread being 42 inches in length, the whole length of thread was therefore = $2800 \times 42 = 117600$ inches; or 3266 yards and two feet, or 260 yards less than 2 English miles." — this gives nearly 48 yards to each grain in weight."

Here the matter rested 'till the year 1784 upon the 24th of november, when being at Munich, where the culture of silk has been introduced under the auspices of the present Elector, I procured a small *cochon* of yellow silk, which being wound singly, the thread was found to be 400 English yards in length; and when thoroughly dried it was found to weigh very exactly $1\frac{1}{2}$ grs. This gives 260 yards to 1 grain in weight; which shows that the raw silk used in the experiment before mentioned, made in the year 1781 must have been composed of at least 5 or 6 single threads wound together.

"Upon the 26th of November, I procured 17 indifferent *cochons* which being wound singly into a skain, the thread was found to be 9051 yards in length; and which being weighed when apparently dry was found to weigh $37\frac{1}{8}$ grains."

"Nov.$^r$ 29. — This silk having now been dried thoroughly in a heat of near 100° of Fahrenheits thermometer, in which it was exposed three days and three nights it was found to weigh but $35\frac{7}{16}$ grs."

"Dec.$^r$ 7th. The same having now lain in water, exposed to a heat of 100°F 48 hours and being thoroughly free'd from the small air-bubbles that adhered to it, was found to weigh *in water* warmed to 65°F $9\frac{13}{16}$ grs, when weighed in the glass bucket. — Upon being carefully taken out of the bucket under water, and weighed loose in the water it weighed $8\frac{1}{2}$ grs only."

Here the same extraordinary diminution of the apparent weight of the silk upon its being taken out of the bucket, and weighed loose in the water, took place as in the experiment before mentioned, and I began now seriously to imagine that I had decovered the cause of it. I concluded that it could arise from no other cause than the adhesion of its parts, or want of fluidity in water: and I was not a little strengthened in this opinion, when, upon calculation [of] the surface of the silk made use of in the last experiment I found it amounted to no less than 600 superficial inches, while the surface of the glass bucket did not amount to much more than 10. — I reasoned thus — The apparent diminution of weight cannot arise from the attraction between the silk and the water alone, for however strong might be the attraction between the particles of water actually in contact with any solid body and the body itself, this could not prevent these particles from sliding by the neighbouring ones; and 'tho these particles might actually remain attached to the body, yet the body would descend notwithstanding, and with the whole force of the difference of the specific Gravities of the body and of the fluid; But if in addition to this attraction between the water and the solid body, there exists an attraction or adhesion of the particles of water among themselves, then, in that case, in order that the body may descend the difference of the specific Gravities must be so great as to overcome this adhesion, and the apparent weight of the body will be diminished in proportion to its surface and to the force, or intensity of this adhesion or attraction.

Now when the silk is weighed in the bucket, the silk, and the bucket, and the contained water, all gravitate together, and the attraction or adhesion among the particles of the water in the vessel cannot hinder the descent of the bucket with its contents but in proportion to the external surface of the bucket; but when the silk is weighed loose in the water this adhesion acting upon the whole surface of the silk, which is incomparably greater than that of the bucket, its effect in diminishing the apparent weight must be greater in the same proportion.

Suppose the adhesion among the particles of water to be equal to 1 grain to a surface of 500 inches, or, $\frac{1}{500}$ part of a grain to each inch: — as the surface of the bucket was only 10 inches the apparent weight of it and its contents could only be diminished $\frac{1}{500} \times 10 = \frac{1}{50}$ part of a grain: but the surface of the silk being $= 600$ inches its apparent weight when weighed loose in the water would be diminished $\frac{1}{500} \times 600 = 1\frac{1}{5}$ grs, consequently the apparent weight of the silk when weighed loose in the water would be less than when weighed in the bucket by $1\frac{1}{5}$ grs — $\frac{1}{50}$ grs $= 1\frac{9}{50}$ grs, and this is nearly the proportion in

which the apparent weight of the silk was actually diminished in both the Experiments before mentioned.

Pursueing this train of reasoning still further I imagined
that if it were possible by further subdivision to increase
the surface of the silk made use of in the last mentioned
experiment eight or nine fold its apparent weight when
weighed loose in water would be totally destroyed, or it
would remain suspended in the water; For, supposing its
surface actually amounted to no more than 500 inches, as
its apparent weight in water was no more than $8\frac{1}{8}$ grs and
as to every 500 inches of additional surface there would
necessarily be a corresponding diminution of apparent
weight equal to 1 grain, if its surface instead of being 500
inches should by any means become $500 \times 8\frac{1}{2} = 4250$
inches its apparent weight would be reduced to nothing.

I went further (for who can restrain the sallies of a
sanguine imagination,) — I conceived the idea of account-
ing for the suspension of the salt in the water of the
Ocean: — the solution and suspension of the earthy particles in the waters of Rivers; — and even the suspension
of the metals in solution in their different menstrua, upon
the same principles — I even went so far as to compute
the number of equal solid spheres into which a grain in
weight of gold must be divided in order to its being suspended in water; — or in other words; in order that its
surface may amount to 500 square inches.

Doctor *Priestley* has found that when pure water and
pure Mercury are agitated together the water loses its
transparency and turns quite black, but that it regains its
transparency immediately upon being heated, a black
powder being precipitated which upon being dried and exposed to a moderate heat assumed the form of pure runing
Mercury. Is it not possible that the blackness of the water

might be owing to the suspension of the extremely small particles into which the mercury was divided, in consequence of the want of perfect fluidity in the water; and that the precipitation of this mercury upon the application of heat and the consequent restoration of the transparency of the water arose from the fluidity of the water being increased or the attraction among its particles being diminished by the heat applied? —

Whether the adhesion or attraction among the particles of fluids, here supposed, actually exists or not; and whatever part it may have in the phenomena attempted to be accounted for; upon this supposition as subsequent experiment convinced me that the appearances upon which all my reasonings upon the subject were originally found arose from a cause totally different from that to which I attributed them, calling in my fancy from these wanderings, I directed my whole attention to the patient and persevering search after truth.

My ballance having been altered and much improved, by Wolf, an excellent Bavarian Artist and rendered sensible to $\frac{1}{500}$ part of grain, I repeated the Experiments, with variations, as follows.

"Ammerlandt, near Munich 15th Sepr. 1785 "Fine clear sun-shine. — Thermometer at 65°F. The raw silk made use of in the Experiment of the 26th of November the last year, (being a single thread 9051 yards in length). being taken out of the drawer belonging to the box containing my Hydrostical ballance, and weighed in air, was found to weigh $32\frac{81}{128}$ grains *Vienna Weight* — N.B. 10000 grains Vienna weight are equal to 11248 Grains Troy."

"This silk being afterwards thoroughly dried at the fire, and being weighed while yet warm was found to weigh

30 Grains, very exactly. — after remaining in the scale 2 minutes its weight was augmented to 30¼ Grains; — and at the end of 4 minutes it was found to weight 30½ Grains. After 3 or 4 minutes more had elapsed it weighed 30¾ — and at the end of about 10 minutes it had augmented in weight *one whole grain*. — This augmentation undoubtedly arose from the moisture attracted from the circumambient air."

"The silk was now put into a large glass jar of clear spring water, at the temperature of the atmosphere, viz. 65°F, and being thoroughly wet it very readily sunk to the bottom of the vessel, where it remained to all appearance totally free of air; but suspecting notwithstanding these appearances that some small invisible air bubbles might still remain behind, attached to the silk, for greater security I placed the vessel in a window, against the light, where I could observe what passed with greater ease and certainty, and suffered it to remain at rest, in order that the water might have time to penetrate the silk more thoroughly, and to drive away the remaining air, if there should be any still behind."

"At the end of about half an hour, examining the silk attentively I was surprised to discover a number of very small air-bubbles attached to its surface, and still more so when I found that they increased very fast, no only in numbers, but in size also. At the end of about three hours and a half, the silk appearing to be covered with them, it 'rose to the surface of the water."

"Having now seperated these air-bubbles from the silk, by shaking it in the water, it sunk again; but it soon began to collect fresh air-bubbles, and at the end of three or four hours it 'rose again, as before, apparently covered with them."

"I detached these air-bubbles a second time, and the silk sunk to the bottom of the vessel as before, but it did not remain there long; Fresh air-bubbles soon begin to make their appearance upon its surface, which increasing in number and size as before it 'rose a third time to the surface of the water."

"I continued to repeat this Experiment four days successively, and always with the same result; when, seeing no prospect of being able to attain of the object of my research by this route, I had recourse to other expedients in order to free the silk of air, and to determine its weight in water."

"September 23ᵈ — The silk being put into a large glazed earthen Vessel with clear spring water, the water was made to boil near two hours, over a brisk fire, after which being removed from the fire, and suffered to cool to the 126° of Fahrenheits Thermommeter, the silk was weighed, *loose in the water*, and was found to weight 10¼ grains at 4ʰ 20′ P.M.;

| | | | |
|---|---|---|---|
| at 4ʰ 30′ P.M. having cooled to 123°F | it weighed | $10\frac{1}{2}$ | Grs |
| at 4ʰ 40′ | 120° | $10\frac{5}{8}$ | |
| at 4ʰ 50′ | 118° | $10\frac{13}{16}$ | |
| at 5ʰ 2′ | 113° | $10\frac{15}{16}$ | |
| at 5ʰ 10′ | 110° | $11\frac{1}{16}$ | |
| at 5ʰ 30′ | 105° | $11\frac{3}{16}$ | |
| at 5ʰ 45′ | 101° | $11\frac{5}{16}$ | |
| at 6ʰ 5′ | 97° | $11\frac{7}{16}$ | |
| at 6ʰ 25′ | 92° | $11\frac{7}{16}$ | |
| at 6ʰ 50′ | 90° | $11\frac{11}{16}$ | |
| at 7ʰ 10′ having cooled to | 87° | it weighed | $11\frac{13}{16}$ Grs |
| at 7ʰ 50′ | 83° | $11\frac{15}{16}$ | |
| at 8ʰ 30′ | 80° | $12\frac{1}{16}$ | |
| at 9ʰ 27′ | $75\frac{1}{2}$° | $12\frac{3}{16}$ | |

and the next morning Sepᵗ 24th at 8 o'Clock the Thermometer in the water standing at 63°, it weighed $12\frac{179}{256}$ grs."

"The silk being now crowded into the small glass bucket, (but without being brought into contact with the air,) and weighed in the water, it was found to weigh $11\frac{13}{32}$ grs."

"Being taken out of the bucket, under water, and suffered to remain loose in the water, at the bottom of the vessel about fifteen minutes, and then being again put into the bucket and weighed in it, in the water, it was found to weigh but $10\frac{7}{16}$ grains."

"Suspecting that this sudden diminution of the apparent weight of the silk might arise from a portion of air communicated to it by my hands in the operation of putting it into the bucket, &c., to ascertain this fact, I now removed it again from the bucket, and handling it very much in the water for about half a minute, I again replaced it in the bucket, and weighed it, when it was found to weigh $8\frac{3}{4}$ grs only."

"Notwithstanding this remarkable diminution of the apparent weight of the silk in the three last Experiments viz. from $12\frac{179}{256}$ grs to $8\frac{3}{4}$ grs, I could discover no air bubbles attached to it 'tho I examined it with the utmost care: and it is certain that no air could have been communicated to it either by the bucket or by the thermometer made use of for ascertaining the temperature of the water, for both these instruments were put into the water at the same time with the silk, that is to say, before the water was boiled; and they were constantly kept in it 'till the end of the Experiment, without once being brought into contact with the air."

*Remarks* upon the foregoing *Experiments*

From these experiments it appears that silk possesses a power of attracting and imbibing water from the air; and that with very considerable force, for in the experiment of the 15th of Sep.ᵣ the silk which being thoroughly dried before the fire, and weighed while yet warm, was found to weigh but 30 Grains, acquired an addition of weight of *one whole grain* upon being exposed 10 minutes in the free air of my room, and this at a time when the air was apparently very dry, and the same silk before it was dried at the fire, was found to weigh $32\frac{81}{128}$ grains; 'tho even in that state it did not appear to be in the least damp.

Hence it appears that a Merchant or manufacturer who purchases 100 lbs. of raw silk, in a common state as to dryness actually pays for at least 8 lbs. of water; and if it has been kept for any considerable time in a damp place the quantity of moisture imbibed will be still greater; and this is surely an object worthy of the attention of those who deal largely in that valuable Commodity.

This property of silk imbibing water from the air is by no means a new discovery, but I believe it has not generally been imagined that the quantity of moisture commonly contained by it is so considerable as by my experiments it appears to be.

Besides this property of attracting water from air; silk possesses another still more extraordinary, which is that of attracting *air* from *water*. This appears evidently from the Experiment of the 15th of Sep.ᵣ 1785, and those made upon the four subsequent days; and by a great number of other Experiments, which I have since made, of which I shall give an account hereafter.

It likewise possesses a power of attracting, and appropriating to itself, air, from the surfaces of other bodies brought into contact with it under water. This appears by the successive and very remarkable diminution of the apparent weight of the silk in water in the Experiments of the 24th of September 1785. and to this cause is doubtless to be attributed the apparent diminution of the weight of the silk in water in the Experiments of the 20th Jan.ʸ 1781, and the 1st December 1784, and not to any adhesion, or want of perfect fluidity among the particles of water, as I then supposed. In the operation of taking the silk out of the glass bucket it doubtless received a small portion of air from my hand, and this rendered it apparently lighter when it was weighed immediately after, loose in the water, than it appeared to be just before, when weighed in the bucket. Upon reversing the Experiment (upon the 24th of Sep.ʳ 1785.) the silk was found to weigh more when weighed loose in the water than when weighed immediately after in the glass bucket, which is a convincing proof that the diminution of the apparent weight of the silk in the former Experiments could not arise from the cause to which I at first attributed it.

But however conclusive the result of these last Experiments may be against the opinion formerly taken up of the cause of the apparent diminution of the weight of the silk under the circumstances described, yet is it not probable that the supposed adhesion, or want of perfect fluidity among the particles of fluids actually exist, at least in some degree? and if it exists must it not necessarily produce effects similar to those attributed to it? — This appears to me to be a matter deserving of investigating, and to which I mean the first leisure time I may have to direct my enquiries. If a solid body can be found whose surface may be very greatly increased without altering its specific gravity, such a body may be made use of for ascertaining the facts in question. Suppose for instance an ounce of gold to be drawn into the finest wire possible, and that this wire is afterwards flatted in order still further to increase its surface, if this wire is exactly counterballanced in air by a solid globe of gold suspended to the opposite arm of a ballance, the question is whether they will remain in equilibrio upon being both immersed in water, care being taken to free them both, and the water itself, of air, by boiling? — If the specific gravity of gold should be found to be altered in the operation of drawing it into wire, (and this appears to be the more probable considering the remarkable change that the specific gravity of Platina undergoes under a similar operation,) Perhaps glass may be used instead of it. It is well known that glass may be drawn out when hot into an exceeding fine thread commonly called *spun Glass*, and that this thread when cold retains its flexibility; and I see no reason to suspect that the specific gravity of Glass is capable of any sensible alteration.

Perhaps in this way the *measure of the fluidity of a fluid* may be ascertained; or, which amount to the same thing, the *measure of its want of perfect fluidity*, and that this may enable us to account for several operations of nature very difficult to be accounted for upon the principles generally received.

But to return to my Experiments upon silk; — of all those which I had made none appeared to me more extraordinary than that of the 23$^d$ of September. That silk which had, as I thought, been thoroughly free'd of air, when weighed in hot water, which had likewise been free'd of air, should increase in weight as the water grew cold, was a paradox, which for a long time I was not able to unravel. If the silk had lost of its apparent weight, as the water lost of its heat, it would have been no more than what I expected, from the condensation of the water with the cold; but its *increase* of weight under these circumstances was no only unexpected, but at first totally unaccountable. I could not imagine that it arose from the condensation of the silk with the cold for this would have been to suppose the expansion of silk with heat incomparibly greater than that of any other known solid body, and even much greater than that of any known fluid air excepted; For the augmentations of its apparent weight from this cause could only be in proportion to the excess of its condensation above that of the water under the given diminutions of heat; and these augmentations appeared to me much too considerable to be accounted for upon this supposition.

At length I concluded that it could arise from no other cause than a portion of air still remaining attached to it (notwithstanding its being so long kept in boiling water) a part of which being afterwards imbibed by the water as it grew cold, and as its power of imbibing air was increased, the silk necessary became apparently heavier in proportion to its loss of air.

I conceived silk in water to be under similar circumstances with respect to the air dissolved in the water, as silk in air is under with respect to the water dissolved in the air. In the one case the silk imbibes the water dissolved in the air, as is well known; and in the other it imbibes or attracts, and appropriates to itself, the air dissolved in the water, as appears by my Experiments; and in the one and in the other case, this can only happen in consequence of its greater affinity with the body attracted, than the affinity of the medium in which it is placed with the same body; and the relative quantities of the dissolved fluid imbibed by the silk, and retained by the medium, will depend upon the relative affinities of the silk, and of the medium with the dissolved fluid in question, and consequently must vary with any cause which produces a variation of these affinities.

Now it is well known that heat increases the affinity of air to water; and It is equally certain that it *diminishes* the affinity of water to the air dissolved in it; but it does by no means follow that it has the same effect upon the silk placed in these fluids, — on the contrary it is very certain that it has not; at least not in the same degree.

If a quantity of silk perfectly dry, be exposed in air, in a middling state as to moisture, and of a constant temperature, in a close room, or in any other confined space; the silk will imbibe a certain quantity of water from the air, (which may be ascertained by its increased of weight), after which it will remain as it were in equilibrio with the surrounding medium without any further augmentation of its apparent weight, or any diminution of it.

If the heat be now increased the silk will be found to lose a part of its weight and will become apparently drier than before, and this not because the quantity of moisture in the confined space is diminished; — for that cannot happen — but because the power of air to imbibe water is more increased by the heat than that of the silk in consequence of which the silk looses a part of its water.

If the heat is diminished the silk will increase in weight; — because the power of the air to imbibe, or hold water, being more diminished than that of the silk with the given diminution of heat, the silk must acquire an augmentation of water before the equilibrium of these powers can be restored.

And hence it appears why hygrometers march towards *dryness* upon any increase of heat in the atmosphere, and towards *moisture* upon any decrease of it, notwithstanding that the quantity of water actually dissolved in the atmosphere remains the same.

Suppose a quantity of silk to be placed in air confined in a close vessel, and suppose the heat to be afterwards diminished to the utmost; or at least 'till the power of the air to imbibe or hold water in solution is totally destroyed or reduced to nothing; in this case the silk will continue to increase in weight 'till it is compleatly saturated with water, after which the water forming upon its surface in drops, will fall off in the form of rain, and the silk will receive no further augmentation of weight.

Suppose now the water thus separated from the air and precipitated from the silk to be taken away, as well as that which will be found attached to the sides of the vessel, and that heat be now gradually applied 'till the air and the silk have acquired their former temperature. The consequence of this will be, that the air, recovering its power of attracting water, as it recovers its heat, will seize upon the water remaining attached to the silk, and appropriating a great part of it to itself, the silk will gradually lose its weight till at length the air having acquired its original temperature, the silk will be found to be much lighter than when the experiment was begun.

Let us now see what would happen if the silk instead of being placed in air, be placed in water, and that the power of the water to hold air in solution be by any means diminished and destroyed in the same manner as we have just supposed with respect to the power of air to hold water in solution. Suppose a quantity of silk to be placed in a vessel of water, and that the water be gradually heated & at length be made to boil, and after boiling for some time that it be suffered gradually to cool 'till it has returned to its original temperature, care being taken to keep the silk under the surface of the water during the whole of this time, to prevent its coming into contact with the atmosphere. According to the principles laid down the following appearances ought to take place. Suppose the

silk at the beginning of the experiment to be apparently
free of air, that is to say, totally free of visible air-bubbles,
and that it reposes quietly at the bottom of the vessel; as
the heat of the water is augmented, its power of holding
the air it contains in solution being diminished in a greater
proportion than the affinity of the silk to the same air is
diminished by the same degree of heat, the air will begin
to quit the water, and to attach itself to the silk, and by
degrees air-bubbles will begin to make their appearence,
which increasing very rapidly, both in number and size,
the whole surface of the silk will very soon appear to be
covered with them.  These air-bubbles rendering the silk
specifically lighter than the water it will attempt to 'rise
to the surface, but being prevented from doing this by
means used for that purpose, the air-bubbles still going on
to increase in size, not only on account of their increas'd
elasticity with their increase of heat, but also in consequence of their continual supply of fresh air furnished by
the water, they will at length break loose from the silk and
rising to the surface of the water will escape.   These
bubbles will be immediately succeeded by others, which
increasing in size in like manner, will detach themselves
and escape in their turns; and this operation will continue
so long as the heat of the water continues to be augmented,
or its power of holding air in solution is successively
diminished.

This operation of the formation and ascent of the air-
bubbles is analagous to the formation and descent of the
drops of water from the silk placed in air super-saturated
with water, above discribed.

During the heating of the water the air separated from
it will not only attach itself to the silk but will also form
bubbles upon the bottom and sides of the vessel, and upon

the surface of any other sort of body that is put into the water, in like manner as drops are formed upon the sides of a vessel containing air supersaturated with water; and these bubbles will begin to detach themselves and escape in vast numbers long before the water has acquired a degree of heat necessary to make it boil.

Suppose the water now to have boiled for a considerable time, and the greater part of the air it contained to have made its escape, and that it is now removed from the fire and suffered to cool. — The silk will still be found to contain a quantity of air for 'tho the heat deprives water of the power of holding air in solution, it does not enable it to deprive other bodies of the power of retaining it; and accordingly we see the sides and bottom of a vessel containing water which has *just ceased to boil,* covered with air-bubbles; and in my Experiments, the silk upon being kept in boiling water above two hours remained constantly covered with air-bubbles during the whole of that time; and even for several minutes after the water was taken from the fire, and had ceased to boil.

These air-bubbles very soon disappear, being imbibed by the water, which regains its power of dissolving air, as it looses of it heat; but from hence it is not to be infered that the silk is left totally free of air any more than that silk in air is totally free of water because there are no longer any visible drops remaining attached to it.

The silk actually retains a portion of air after all the visible air-bubbles have disappeared; but the power of the water to imbibe it continuing to increase as its heat is diminished, the quantity of air retained by the silk will be continually lessened, and *consequently its apparent weight in the water will be continually augmented as the water goes on to cool*; agreeable to the result of the result of the experiment before mentioned.

I have been the more particular in my indeavours to account for this Phenomenon as it is a new appearance, (at least I do not recollect to have seen it noticed by any body,) and as it is of importance to be known to those who make Experiments upon certain bodies with a view to determine their specific Gravities.

And this brings me back to the *specific gravity of silk,* the original object of these my Experiments.

Its specific gravity as determined by the Experiments made in the year 1781, turned out to be to that of water as 1337 to 1000, but it appears by the Experiments of the last year, (1785) that this decision was very far from being accurate: for taking the weight of the silk used in these last mentioned Experiments, in air = 30 Grains, as found in the Experiment of the 15th of September; and its weight in water = $12\frac{179}{256}$ Grains, as found upon the morning of the 24th of September, its Specific Gravity turns out to be that of water as 1734 to 1000; and there is reason to believe that this is very near the truth.

The silk made use of in these Experiments consisted of a single thread, as spun by the worm. 9051 English yards in length, weighing 30 Grains Vienna weight = 33,744 Grains Troy; and it lost of its weight upon being weighed in water $17\frac{77}{256}$ Grains Vienna weight, or $19,46$ Grains Troy, consequently its volume was equal to the volume of $19,46$ Grains Troy of water; and as an English cubic inch of water weighs 253,$185$ Grains Troy the volume of a quantity of water weighing $19,46$ Grains Troy, and consequently the volume or solid contents of the silk is $\frac{253,185}{19,46} = 0,076811$ of a cubic inch.

Now as the length of the thread of silk was 9051 yards = 325826 English Inches, the area of its transverse section must have been no more than $\frac{0,0768611}{325826} = \frac{1}{4239153}$ of an inch; or it would require *four million, two hundred and thirty nine thousand, one hundred and fifty three* threads of raw silk to form a solid rope or cylinder the area of whose transverse section should be *one inch;* — which is a degree of fineness really astonishing.

The weight of the silk being 33,744 Grains, and the length of the thread 9051 yards, it would require $268\frac{1}{4}$ yards nearly to weigh 1 grain; or $6\frac{26}{100}$ Grains in weight of this thread would be 1 English mile in length. This gives more than 73 miles in length to a single thread weighing 1 ounce Troy; Consequently if a Silk Gown worn by a lady weighs 28 ounces it is very certain that she carries upon her back upwards of 2000 miles in length of silk, as spun by the worm. — and hence it appears that a man might actually carry in his Pockets a thread long enough to reach round the world; — for a thread of raw silk 25000 miles in length, (which is nearly the circumference of the Globe), would weigh no more than $341\frac{3}{4}$ ounces. And a quantity of silk sufficient to load an English broad-wheeled-wagon drawn by eight Horses would contain a length of thread so great that a Canon Bullet flying continually day and night with the same rapidity with which it leaves the mouth of the Piece would require more than *fourteen years* to pass from one end of it to the other.

But what is still more extraordinary, if possible, is, that this thread 'tho so extremely fine is perfectly visible, even to the naked eye, and that at the distances of several feet. The area of its transverse section being $= \frac{1}{4239153}$ part of an inch its diameter computed from thence turns out to be $= \frac{1}{1824}$ part of an inch very nearly and the thread being visible, as we may conclude that a piece of the thread equal in length to its diameter would be visible likewise, it appears that a body whose volume is no more than

$= \frac{1}{7735095353}$ part of a cubic inch is discernable to the naked eye, or if instead of this little cylinder of raw silk we suppose a globe to be formed of the same diameter, the volume of this globe will be less than *one ten thousand millionth part of an inch*.

The diameter of the thread being $= \frac{1}{1824}$ of an inch if it be placed at the distance of four feet from the eye it will subtend an angle less than $\frac{1}{25}$ part of a minute or $2'' . : 20'''$, and at that distance it is perfectly visible to a good eye, in a moderately enlightened room, upon what-ever colored ground it be placed.

If it be placed against a back ground of black velvet at the distance of 4 or 5 inches from it, and if the day be clear it will be visible to a sharp eye at the distance of 30 feet, tho' at this distance the thread cannot subtend an angle greater than about $\frac{1}{3}$ part of a second, or $18''' . . 52''''$

When the thread is inlighten'd by the direct rays of the sun falling upon it, it is visible at a still greater distance; but at its diameter at the same time appears to be much greater than it really is, no computation for ascertaining the angle a body must subtend in order to its being visible can with any certainty be founded upon this Experiment.

To determine the weight of the smallest visible particle of silk, say as 268,225 yards $=$ 9656 inches of the thread weigh no more than 1 grain. This gives $\frac{1}{9656}$ part of a grain to each inch, and as we supposed $\frac{1}{1824}$ part of an inch in length, of the thread, to be visible, the weight of this little cylinder will be $\frac{1}{9656} \times \frac{1}{1824} = \frac{1}{17612544}$ part of a Grain; — or reducing this cylinder to a globe of the same diameter, its weight will be no more than $\frac{1}{26418816}$ part of a grain.

But I do not pretend to lay any great stress upon these computations respecting the dimensions of the least bodies visible, and their weights; for it does not seem to be quite certain that because a thread of raw silk is visible to the naked eye that a globe of the same diameter would be visible likewise.

In my next letter I shall send you an account of the Experiments which I have made upon the air separated from water, by silk; under various circumstances, together with an account of several other experiments relative to the air yielded by water.

In the meantime I have the honor to be

&c. —

# AN ACCOUNT

OF SOME

# EXPERIMENTS

MADE

## TO DETERMINE THE QUANTITIES OF MOISTURE ABSORBED FROM THE ATMOSPHERE BY VARIOUS SUBSTANCES.

**B**EING engaged in a course of experiments upon the conducting powers of various bodies with respect to heat, and particularly of such substances as are commonly made use of for cloathing, in order to see if I could discover any relation between the conducting powers of those substances and their power of absorbing moisture from the atmosphere, I made the following experiments.

Having provided a quantity of each of the undermentioned substances, in a state of the most perfect cleanness and purity, I exposed them, spread out upon clean china plates, twenty-four hours in the dry air of a very warm room (which had been heated every day for several months by a German stove), the last six hours the heat being kept up to 85° of Fahrenheit's thermometer; after which I entered the room with a very accurate balance, and weighed equal quantities of these various substances, as expressed in the following table.

This being done, and each substance being equally spread out upon a very clean china plate, they were removed into a very large uninhabited room upon the second floor, where they were exposed 48 hours upon a table placed in the middle of the room, the air of the

455

room being at the temperature of 45° F.; after which they were carefully weighed (in the room), and were found to weigh as under mentioned.

They were then removed into a very damp cellar, and placed upon a table in the middle of a vault, where the air, which appeared by the hygrometer to be completely saturated with moisture, was at the temperature of 45° F.; and in this situation they were suffered to remain three days and three nights, the vault being hung round, during all this time, with wet linen cloths, to render the air as damp as possible, and the door of the vault being shut.

At the end of the three days I entered the vault with the balance, and weighed the various substances upon the spot, when they were found to weigh as is expressed in the third column of the following table : —

| The various Substances. | Weight after being dried 24 hours in a hot room. | Weight after being exposed 48 hours in a cold uninhabited room. | Weight after being exposed 72 hours in a damp cellar. |
|---|---|---|---|
| | Pts. | Pts. | Pts. |
| Sheep's wool   .   .   . | 1000 | 1084 | 1163 |
| Beaver's fur   .   .   . | 1000 | 1072 | 1125 |
| The fur of a Russian hare   . | 1000 | 1065 | 1115 |
| Eider-down   .   .   . | 1000 | 1067 | 1112 |
| Silk { Raw, single thread   . | 1000 | 1057 | 1107 |
| Silk { Ravelings of white taffety | 1000 | 1054 | 1103 |
| Linen { Fine lint   .   .   . | 1000 | 1046 | 1102 |
| Linen { Ravelings of fine linen | 1000 | 1044 | 1082 |
| Cotton-wool .   .   .   . | 1000 | 1043 | 1089 |
| Silver wire, very fine, gilt, and flatted, being the ravelings of gold lace   .   . | 1000 | 1000 | 1000 |

N. B. The weight made use of in these experiments was that of Cologne, the *parts*, or least divisions, being $= \frac{1}{65536}$ part of a mark; consequently 1000 of these *parts* make about $52\frac{3}{4}$ *grains Troy*.

I did not add the silver wire to the bodies above mentioned, from any idea that that substance could possibly imbibe moisture from the atmosphere; but I was willing to see whether a metal, placed in air saturated with water, is not capable of receiving a small addition of weight from the moisture attracted by it, and attached to its surface; from the result of the experiment, however, it should seem that no such attraction subsists between the metal I made use of, and the watery vapour dissolved in air.

I was totally mistaken in my conjectures relative to the results of the experiments with the other substances. As linen is known to attract water with so much avidity; and as, on the contrary, wool, hair, feathers, and other like animal substances are made wet with so much difficulty, I had little doubt but that linen would be found to attract moisture from the atmosphere with much greater force than any of those substances; and that, under similar circumstances, it would be found to contain much more water; and I was much confirmed in this opinion upon recollecting the great difference in the apparent dampness of linen and of woollen clothes, when they are both exposed to the same atmosphere. But these experiments have convinced me that all my speculations were founded upon erroneous principles.

It should seem that those bodies which are the most easily wetted, or which receive water, in its unelastic form, with the greatest ease, are not those which in all cases attract the watery vapour dissolved in the air with the greatest force.

Perhaps the apparent dampness of linen to the touch arises more from the ease with which that substance parts with the water it contains than from the quantity of water it actually holds; in the same manner as a body appears hot to the touch, in consequence of its parting freely with its heat, while another body, which is actually at the same temperature, but which withholds its heat with greater obstinacy, affects the sense of feeling much less violently.

It is well known that woollen clothes, such as flannels, &c., worn next the skin, greatly promote insensible perspiration. May not this arise principally from the strong attraction which subsists between wool and the watery vapour which is continually issuing from the human body?

That it does not depend entirely upon the warmth of that covering is evident; for the same degree of warmth, produced by wearing more cloathing of a different kind, does not produce the same effect.

The perspiration of the human body being absorbed by a covering of flannel, it is immediately distributed through the whole thickness of that substance, and by that means exposed by a very large surface to be carried off by the atmosphere; and the loss of this watery vapour, which the flannel sustains on the one side, by evaporation, being immediately restored from the other, in consequence of the strong attraction between the flannel and this vapour, the pores of the skin are disencumbered, and they are continually surrounded by a dry, warm, and salubrious atmosphere.

I am astonished that the custom of wearing flannel next the skin should not have prevailed more universally. I am confident it would prevent a multitude of diseases; and I know of no greater luxury than the comfortable sensation which arises from wearing it, especially after one is a little accustomed to it.

It is a mistaken notion that it is too warm a cloathing for summer. I have worn it in the hottest climates, and in all seasons of the year, and never found the least inconvenience from it. It is the warm bath of a perspiration confined by a linen shirt, wet with sweat, which renders the summer heats of the tropical climates so insupportable; but flannel promotes perspiration, and favours its evaporation; and evaporation, as is well known, produces positive cold.

I first began to wear flannel, not from any knowledge which I had of its properties, but merely upon the recommendation of a very able physician (Sir Richard Jebb); and when I began the experiments of which I have here given an account, I little thought of discovering the physical cause of the good effects which I had experienced from it; nor had I the most distant idea of mentioning the circumstance. I shall be happy, however, if what I have said or done upon the subject should induce others to make a trial of what I have so long experienced with the greatest advantage, and which, I am confident, they will find to contribute greatly to health, and consequently to all the other comforts and enjoyments of life.

I shall then think these experiments, trifling as they may appear, by far the most fortunate and the most important ones I have ever made.

With regard to the original object of these experiments, the discovery of the relation which l thought might possibly subsist between the warmth of the substances in question, when made use of as cloathing, and their powers of attracting moisture from the atmosphere; or, in other words, between the quantities of water they contain, and their conducting powers with regard to heat, I could not find that these properties depended in any manner upon, or were in any way connected with, each other.

# EXPERIMENTS AND OBSERVATIONS

## ON THE

## ADVANTAGE OF EMPLOYING WHEELS WITH BROAD FELLOES FOR TRAVELLING AND PLEASURE CARRIAGES.

---

WHEN we consider the immense number of coaches, diligences, cabs, and other vehicles, for travelling or for pleasure, which are to-day in use among the various nations, and the great number of horses employed in drawing them, we shall see that every improvement in the construction of these carriages, which without being too expensive renders them either more agreeable or more durable or easier to draw, would deserve to be considered an object of very great importance to society, and consequently well worthy the attention of those who love to contribute to perfecting useful things.

As far as the preservation of the roads is concerned, no one has ever doubted the advantages to be gained by the adoption of the wheels with broad felloes which have been prescribed for some years in France, for large wagons and other vehicles intended to carry heavy loads ; but opinions have been divided on the question, whether these new wheels did not make the wagons heavier, and harder to draw. Experience has rapidly scattered the fears of the wagoners in this

461

respect; but the people at large, always slow in all
countries to interest themselves in novelties which
have only their utility to recommend them, are still
very far from suspecting the great advantages which
must result in the end from this change in the construction of wheels, when it is generally adopted for all
sorts of vehicles, as it can hardly fail to be sooner or
later in all countries where roads are well finished.

As long as the roads were bad and the ruts deep, it
was impossible to use any wheels except those with
narrow felloes; but, now that there are good roads
almost everywhere, one cannot long avoid the conviction that wheels with broad felloes are preferable to
others, especially when they are intended for use on a
paved road.

If we watch carefully the wheel of a carriage which
is being drawn over a paved road, we shall see that it
is tossed about very much, slipping continually to the
right and left, falling into all the spaces between the
stones, and then striking roughly against the stone
immediately before it. These sharp blows, following
one another rapidly, give very disagreeable shocks to
the carriage, and strain the wheels so that they soon
wear out. They strain the carriage still more, and
affect the horses by giving them severe jerks, and
make the draught unequal and very toilsome. Nor
does the evil end here: the tires, although flat when
new, are soon rounded at their edges by this continual
slipping right and left, so that the wheels, if narrow,
become every day more inclined to slip; the stones of
the pavement itself, in the course of time, become worn
and rounded; the spaces between them become wider
and deeper; the wheels fall into these holes more

easily and with greater force, and soon the roads are entirely worn out.

The remedy for all these inconveniences is so simple and so easily found that it is really astonishing that the use of it has been for so long a time neglected.

Struck by the advantages which ought to result from the adoption of wheels with broad felloes for pleasure carriages, I persuaded a person of my acquaintance in Paris, six years ago, to have a pair of wheels for a fly made with felloes 4 inches broad. These wheels were made, under my direction, by M. Groux, a wheelwright living on the Rue de Sèvres; but circumstances, which need not be mentioned here, have always prevented an experiment being made with them.

In the course of a journey to Bavaria, last autumn, I had on the way an opportunity of speaking to several wagoners, whom I met with large wagons carrying heavy loads between Paris and Strasburg; and I learned from them how well they are pleased now with the change which the law has obliged them to make in the construction of the wheels of their wagons. Several of them assured me that, with the same number of horses, they could now load their teams with a load a quarter heavier than they carried formerly with narrow wheels, and that the new wheels are much stronger and more durable than the old ones.

This information strengthened me in the opinion which I had for a long time entertained on the preference which should be given to wheels with broad felloes for all sorts of carriages: and I made on the spot a firm resolve to brave the ridicule which is always encountered by those who dare to be the first to deviate from customs which are consecrated by fashion ; and, on my return to Paris, I had made for my carriage wheels with broad felloes.   I have now for two months used them daily, and I am so well pleased with them that I feel it to be a duty to make known the results of this experiment.   The carriage, which is a two-seated coach, has become incomparably more comfortable and more agreeable than it ever was before ; and I have just discovered, by comparative experiments of which I will give an account, that it has become more easy to draw, and that it is less tiresome for the horses.

Having kept the old wheels, which are not worn out, and also by a happy chance a still older set, which are yet narrower, I had my carriage arranged in such a manner as to be able to measure exactly the force employed by the horses in drawing it ; and, using the three kinds of wheels alternately, always going over the same road at the same rate of speed, and with the same amount of load, I have been able to determine, in a perfectly decisive manner, not only which of the wheels roll the easiest, but also in every case *how much* less is the force exerted in drawing with one set than with the others.

The method by which I estimated the force employed was as follows : A bar of beech-wood, 29 inches long, 4 inches wide, and 1 inch thick, moving without sensible friction in a groove, is placed flat upon the forward axle of the carriage, in the direction in which it is to travel.   At the two ends of this bar of wood are two iron hooks.   To the hook in front is fastened a splinter-bar, and to the ends of this bar the whippletrees are attached.   To the other hook is fastened the end of a stout rope, the other end of which is fixed to a pulley, 3 inches in diameter. This pulley is placed flat upon the forward axle of the carriage, behind the bar of wood above mentioned; so that, when the rope is stretched by the pulling of the horses, it lies in the direction in which the carriage is going.

On the small wooden wheel, three quarters of an inch in thickness, which forms this pulley, another wheel, not quite so thick and 12 inches in diameter, is fixed in such a way that the two wheels, attached the one to the other, form but a single body, turning freely on an iron pivot between two pieces of oak, which are fastened by iron pins to the forward axle of the carriage. A rope, less stout than the first, is fastened at one end to the larger wheel of this double pulley, and encircles it (in an opposite direction, however, to that of the larger rope, which is around the small wheel); and its other end being fastened to the hook of a steel-yard or circular spring-balance, the elasticity of this spring opposes the effort of the horses to draw the carriage, and balances it continually, and the needle of the balance indicates the amount of force employed.

Since the diameters of the two wheels, around which the two ropes pass in opposite directions, are in the proportion of 1 to 4, it is evident that the amount of force indicated by the needle is only one quarter of that put forth by the horses.

The balance which I use is made to weigh 150 pounds: it is therefore evident that it ought to be able to resist the force exerted by the horses in drawing, until this force becomes equal to a weight of 600 pounds; but in the experiments that I have made, up to the present time, that force has never exceeded 300 or 400 pounds, even in the jerks given to the carriage by the horses in shying (which all the care of the driver could not always hinder), nor in the shocks caused by obstacles met by the wheels.

Since the motion of a horse is never perfectly uniform, the force exerted by the horses in drawing a carriage must of necessity vary at every step. This causes the needle, which indicates at any moment the force actually employed at that moment, to oscillate continually, and sometimes with such rapidity that the eye can scarcely follow it. However, notwithstanding this continual oscillation, it is not difficult in ordinary cases to determine with sufficient accuracy the mean force of traction. We have only to take what seems to be the mean between all the oscillations; leaving out of account those which are the result of the shying of the horses, as well as those which are caused by foreign objects, as bits of stones, etc., which the wheels sometimes encounter on all roads.

In order to make this paper more satisfactory and more useful, I must give a detailed description of the different kinds of wheels used in my experiments.

The wheels of my carriage which I had next before the last were made in Munich. They are very light, and very much worn. Their tires, which were originally an inch and three quarters broad, are so much worn and rounded at their edges that it is difficult to say how broad they really are now; and this causes the wheels to slip continually, especially on a worn pavement. I have used them but little in my experiments, for fear they would crush under the weight of the carriage.

My last wheels were made in Paris, by a very skilful workman (M. Garnier, living on the Rue Neuve-des-Mathurins). I have had them already more than two years ; and, although they have been used a great deal, and have made long journeys, they are still in very good condition.   They are broader than ordinary carriage wheels : the tires are two and a quarter inches in width ; and the felloes are wide in proportion, and strong.

My new wheels (also made by M. Garnier) have tires 4 inches broad and 5 lines thick.   The felloes are 4 inches wide, but they are not so thick as those of my last wheels : and, since the spokes are also of less thickness, although somewhat broader, the new wheels, seen from one side, appear lighter and more elegant than the last ones.

The three sets of wheels are of about equal heights. Their several dimensions and weights are as follows : —

|  | Next to the last wheels. | Last wheels. | New wheels. |
|---|---|---|---|
|  | ft. in. lines. | ft. in. lines. | ft. in. lines. |
| Height of front wheels  . . . . | 3  4  0 | 3  2  3 | 3  3  3 |
| ,,   ,,  hind wheels  . . . . | 4  9  3 | 4  8  9 | 4  8  3 |
| Breadth of tires . . . . . . . . | 0  1  9 | 0  2  3 | 0  4  0 |
|  | lbs. | lbs. | lbs. |
| Weight of front wheels  . . . . | 124 | 174 | 240 |
| ,,   ,,  hind wheels  . . . . | 226 | 258 | 360 |
| ,,   ,,  the four wheels  . . . | 350 | 432 | 600 |

|  |  |
|---|---|
|  | lbs. |
| The carriage on the new wheels weighs . . . . . . . . . | 1721 |
| In the experiments made with these wheels, it was loaded with three men, — the owner, the coachman, and the footman, — weighing together   . . . . . . . . . . | 400 |
| So that the total weight drawn by the horses was   . . . | 2121 |

When experiments were made with the old wheels, care was taken to load the carriage with an additional weight, equal to the difference between that of the new wheels and that of the old wheels then employed.   I found however, in the end, that without this addition to the load, made to equalize the weight, the force necessary to draw the carriage was always less with the broad wheels than with the narrow ones, in spite of the fact that the latter were lighter.

This difference of weight was compensated in such a degree by the greater breadth of the wheels, that I think I can assert that the carriage, passing over the paved road on the new wheels, and loaded with two persons besides the coachman and the footman, draws easier, and tires the horses less, than when, on wheels of the breadth of ordinary carriage wheels, it is going empty over the same road at the same speed. It may be judged from this how much I must be impressed with the importance of the subject on which I have endeavoured to throw light.

For the satisfaction of those who desire to know more in detail the results of my experiments, I will give here a copy of the register that I kept when they were made.

On the highway to Versailles, between the Pont de Sèvres and Passy, on the pavement: —

The force exerted in drawing was in pounds.

|  | At a slow walk. | At a fast walk. | At a slow trot. | At a fast trot. |
|---|---|---|---|---|
| With the new wheels . | 40 to 44 | 48 to 56 | 74 to 84 | 120 to 130 |
| With the last wheels . | 44 to 48 | 56 to 60 | 84 to 96 | 130 to 140 |
| With the wheels next before the last . . . | 48 to 60 | 60 to 72 | 96 to 120 | 140 to 150 |

On the same route, on the unpaved road by the side of the pavement, the amount of force varied at each moment, according as the road was more or less sandy. When the road was very good and but little sandy, it amounted to: —

|  | At a slow walk. | At a fast walk. | At a slow trot. | At a fast trot. |
|---|---|---|---|---|
| With the new wheels . . | 76 to 84 | 80 to 84 | 80 to 88 | 80 to 88 |
| With the last wheels . . | 80 to 92 | 80 to 96 | 82 to 100 | 82 to 100 |

For a portion of the distance where the road was rather sandy, at a walk the force amounted to from 92 to 100 pounds with the new wheels, and from 100 to 120 with the old ones; at a trot it amounted to from 100 to 110 pounds with the new wheels, and to from 120 to 130 with the old ones.

Over a part of this road which was still more sandy, the force at a walk, as well as at a trot, was from 120 to 130 pounds with the new wheels, and from 125 to 135 with the old ones.

On a part of the road which was very sandy, the force was from 160 to 180 pounds with the new wheels, and from 180 to 200 with the old ones, at a walk and also at a trot.

On the fine road to Saint Cloud (which is not paved), between the Pont de Saint Cloud and the road to Versailles, the force of traction was, at a walk, from 72 to 80 pounds with the new wheels, and from 80 to 85 with the old ones. At a trot, the force of traction was with the new wheels from 80 to 84 pounds, and with the old ones from 82 to 88.

Over stones recently laid, and on which no carriage had travelled, — on the new road which extends across the fields from Passy to Auteuil — the force at a slow walk, with the new wheels, was from 200 to 240 pounds, and with the old ones from 220 to 280.

In the deepest sand that I could find in the Bois de Boulogne, the force at a slow walk was, with the new wheels, 240 pounds, and, with the old ones, from 260 to 280.

When ascending slowly, by the paved road, the hill which one meets in coming from the high-road to Versailles, just before entering the village of Auteuil, the force was 140 pounds with the new wheels, and 150 with the old ones.

A very remarkable circumstance in the results of these experiments, and one which seems to me sufficiently important to deserve to be generally recognized, is the great effect which the nature of the road has upon the relation which the required increase of force bears to any increase of speed.

We have seen that, when the coach was going at a slow walk over a paved road, the force with the new wheels was only about 40 pounds; but that, at a slow trot, it became equal to 80 pounds, and at a rapid trot it equalled 120 pounds. On an unpaved road, however, as well as in sand, the force was always the same, or very nearly so, whatever the speed of the horses might be. This difference no doubt arises from the severe shocks which the carriage receives when it is drawn rapidly over a pavement; for it is evident that, for each blow which the carriage receives from the stones of the pavement, there is a certain amount of force employed, and this must always be supplied by the horses. From this fact we may draw the important conclusion that, the easier a carriage is to ride in, the less is the force necessary to draw it, its weight and load remaining the same; and, as no one can doubt that wheels with broad felloes must roll over a pavement more easily than narrow wheels, this fact alone is enough to show that they are preferable to the old kind of wheels for all sorts of carriages.

A knowledge of the remarkable fact that the amount of force required to draw a carriage over an unpaved road is not sensibly increased by increasing the speed might be put into practice with advantage on many occasions in husbanding the strength of the horses. It might, in the first place, be the means of deciding the question often agitated, whether, in performing a long journey with the same horses, we ought to follow the example of the Italian *vetturini*, who, starting at day-break, travel the whole day at a walk ; or whether it would not be less tiresome for the horses to travel more rapidly four or five hours each day, and then rest longer in the stable.

During a journey which I made in Italy, in 1793 and 1794, with my own horses, I made some experi-ments to settle this question ; and I found, in fact, that my horses were in a much better condition after trav-elling fifteen days, going eight or ten leagues a day at a trot, than after travelling for the same length of time, and going over the same distance, at a walk. I am now able to give a satisfactory explanation of this result.

Those who have travelled in Italy with post-horses know that the Italian postilions always make their horses gallop when they have to ascend a hill, and that they do not stop galloping until they have reached the top.

As, in this case, the force expended in drawing the carriage is not sensibly greater when going fast than when going slowly, the Italian postilions are perhaps right in trying to pass rapidly over a disagreeable por-tion of the road which they cannot avoid ; and I am so fully convinced of it that I shall not fail to adopt their method in future, and especially in passing quickly over all the small, very sandy portions of the road that I encounter.

If, when travelling on a paved road, one wishes to go very fast, it is better to leave the pavement, and

travel on the unpaved part of the road at the side, even
when this portion of the road is far from being good;
but if travelling with a heavily loaded carriage, and
desiring to spare the horses, it is better to proceed at
a walk on the pavement.

I will conclude this paper with some remarks on the
various objections which might be brought forward to
the adoption of wheels with broad felloes in pleasure
carriages.

It may, perhaps, be said that these wheels must of
necessity be heavier than  ordinary carriage-wheels.
This remark has already been made to me many
times, and this is the reply that I have always given:
It is not absolutely necessary that the wheels with
broad felloes should be heavier than ordinary wheels;
for the hubs and spokes can, without any inconvenience, be of the same dimensions as have up to this
time been given to ordinary wheels; and as far as the
felloes and tires are concerned, if they are made broader,
they may be made thinner, and still, by their very construction, the new wheels will be both stronger and more
durable than wheels of the ordinary form and proportions, having the same weight and the same height.
Since, however, wheels with broad felloes are most
certainly easier to draw than the old-fashioned wheels,
I would always advise making them a little stouter, that
they may be a great deal more durable; and they may
have this additional strength, without its injuring at all
the elegance of shape of the wheel.

If the spokes are made broad, they need not be
made so thick; and this will give them the appearance
of being lighter, especially when the wheel is seen
from one side, and this is the only position of a wheel

in which one can judge of the elegance of its form.
Besides, I am of the opinion that the shape of wheels
with broad felloes is more noble and beautiful than
that of ordinary wheels; and that a painter of good
taste would give it the preference, if he were about
to introduce a chariot or a modern carriage into a
large painting.

Some persons have supposed that wheels with broad
felloes must be harder to draw than ordinary wheels, on
unpaved roads, especially in mud, on account of their
greater adhesion to the road; but the resistance due to
the adhesion which one body experiences in rolling on
another is always so inconsiderable that, in the case in
question, the supposed difference would be altogether
insensible. The resistance arising from the friction of
two bodies sliding one over the other is an altogether
different affair; but I have already shown that the
broad wheels slide less on the road than the narrow
ones.

Others have supposed that broad wheels must take
up more mud than narrow ones: but this supposition
is scarcely better founded than the preceding one; for
the quantity of mud that a wheel can take up must be
in proportion to the amount of surface by which it
comes in contact with the mud. Now, the broader the
felloes of a wheel, the less it sinks into the mud: consequently, a broad wheel ought not to come in contact
with the mud by a larger surface than a narrow wheel
does; it is even very probable that the surface of contact is smaller.

As to the advantage of wheels with large felloes on
the score of economy, they ought assuredly to be superior to the old style of wheels; for, although they may

cost about a quarter more than the latter, as they will last at least twice as long, and require much less repairing, they will be less expensive in the long run.

The tires of the new wheels being twice as broad as those of ordinary coach-wheels, they are much less weakened by the holes pierced to receive the nails or iron pins which fasten them to the felloes : they are, consequently, much stronger and less liable to be broken in use.

As the tires are broad enough to prevent the wheels getting into the spaces between the paving-stones, they will be less worn, and worn more evenly, than the tires of narrow wheels. They will also wear the pavement much less, and do less damage to unpaved roads, and indeed to any sort of road.

It is only necessary to take care that the axle of these new wheels is straight, or nearly so, that these wheels may roll flat upon the road ; for, without this precaution, the wheel will be impeded in its motion, and the tire will be worn more on one side than on the other.

Having had a new axle made for my carriage (5 inches longer than the old one), I have given my new wheels an inclination of only three lines, and that seems to me to be enough.

If, in the case of a carriage provided with ordinary wheels having a good deal of inclination, it is desired to substitute for these wheels others with broad felloes, without changing the axle, it can be done ; but in this case it will be indispensably necessary for the tires of the new wheels to be slightly conical, instead of being cylindrical as they ordinarily are made, and for the felloes to be made of the proper shape to receive them.

I know very well that wheels with conical felloes or tires have one disadvantage; for I was present at the ingenious experiments of Mr. Cummings, which made the fact evident. (See Annales des Arts et Manufactures, Vol. V., p. 88.) This disadvantage, however, — that of grinding the road, — would be hardly sensible in wheels 4 feet high, with felloes only 4 inches broad.

A carriage set on wheels with broad felloes, which turn on a nearly straight axle, will be much less liable to be overturned than ordinary carriages; and this is assuredly a very important advantage, especially in a travelling carriage. Nor, on the other hand, will the carriage be more likely to get locked with another, on account of this change; for the considerable inclination which is now given to the hind wheels causes these wheels to be farther apart above than the new wheels on a suitable axle would be.

As to the exact width which would be the most advantageous for wheels intended for pleasure carriages, that experiment alone can determine. It will be necessary to find it by trial, as I have sought to do. I know for a certainty that wheels 4 inches broad are preferable, in all respects, to those which are only $2\frac{1}{4}$ inches in breadth; but it is quite possible that a carriage mounted on wheels $3\frac{1}{2}$ inches in breadth would be as easy, or almost as easy, as mine on my new wheels.

As long as the tires are broad enough to prevent the wheels sliding from side to side, and tumbling into the spaces between the stones of the pavement, the carriage will roll very easily.

I found that my carriage became perceptibly easier with my last wheels, which were $2\frac{1}{4}$ inches wide, than it had ever been with the preceding ones, which were

$1\frac{3}{4}$ inches wide ; but with the new wheels it has become easy to a degree truly remarkable. I could call to witness several persons who have tried it, and, among others, certain members of the Institute, who are here present.

The carriage, mounted on its new wheels, and having in place the apparatus which I used to measure the force of traction in my experiments, is at the present moment in the court of the Palais de l'Institut : where it will remain for some time after the close of the session, that all who are curious to see it may examine it.

I should have much satisfaction in learning that my labours on this interesting subject have met with the approbation of this illustrious assembly, and that they have judged it worthy the attention of those who have the means of making it useful.

# PLANS FOR THE CONSTRUCTION OF A FRIGATE

## EXTRACT FROM STALKARTT'S NAVAL ARCHITECTURE.

[Naval Architecture, or the Rudiments and Rules of Ship Building. Exemplified in a Series of Draughts and Plans, with Observations extending to the further Improvement of that important Art. Dedicated by permission to his Majesty, by Marmaduke Stalkartt. London: Printed for the Author and sold by J. Boydell, Cheapside, J. Dodsley, Pall Mall, and J. Sewell, Cornhill. 1781. folio. pp. 231.]

### BOOK VII. — INTRODUCTION.

SINCE the former part of the treatise has been in the press, a gentleman, whose eminent talents have called him into the service of government in one of its important offices, has communicated to me his studies and ideas on that subject, in the construction of the draught hereto annexed: they are so similar to my own, and tend so much to corroborate the doctrine which I have laid down, that I thankfully embrace the liberty he has given me of inserting it, and consider myself as fortunate in the acquisition, since to the philosophical conclusions of Mr. Thompson there is joined the practical experience of some of the most distinguished artists in the kingdom. The warm approbation which it has received from these gentlemen, as well as from some of the oldest and best officers in the navy, cannot fail of giving confidence to the student, and of recommending the principle to the attention of the state, by which, it is humbly hoped, it will be reduced to the test of experiment.

477

## OF THE FRIGATE.

Copy *of a* Letter *from* Benjamin Thompson, *Esq.*,
F.R.S., *to Mr.* Marmaduke Stalkartt.

Sir, — Agreeably to your request, I herewith send
you my draught of a frigate, upon a new construction,
which you will make any use of you may think proper.
Though I have little doubt with respect to the principles upon which this drawing is made, yet I should
hardly have ventured to have proposed it to have been
carried into execution; nor should I now have con-
sented to its being made public, had it not been for
the very flattering approbation it has met with from
some of the best judges of Naval Architecture in this
kingdom.

That curious and most important art has long been
my favourite study; and several sea-voyages, particularly
a three months' cruise in the Channel fleet, under the
command of the late Sir Charles Hardy, in the year
1779, afforded me an opportunity of making many
remarks upon the qualities of ships, which in all prob-
ability would not otherwise have occurred to me. It
was during this cruise that I amused myself with making the drawing which I now send you; and, when I
began it, I had little more than amusement in view.
But, after it was finished, it was so much approved
of by many able and experienced seamen to whom I
showed it, that I could not refuse the pressing solici-
tation that was made to me to offer it to the Surveyors
of the Navy, to have a ship built after it, by way of an
experiment; and several officers of rank in the Navy,
and high in the estimation of the profession, voluntarily

engaged to do every thing in their power to get the measure adopted.

I confess, I never had very sanguine hopes of our being able to carry this point. Professional men are seldom disposed to allow others to meddle in their business ; but, thus recommended, I thought it rather probable that we should succeed, but it turned out otherwise.

Having failed in this attempt, I afterwards endeavoured to get the plan carried into execution by private subscription, and several of my friends offered to subscribe very generously for that purpose ; but so large a sum of money was wanted, and so great a length of time was necessary in order to complete the undertaking, that these circumstances, added to the uncertainty of the continuance of the war, prevented my being able to accomplish my design. By the copy of my proposals, which accompanies the draught, you will see the grounds upon which I proceeded in this business ; and, by the certificates annexed to those proposals, you will see the manner in which I was supported. With such respectable testimonies in favour of the plan, I think I cannot risk much in allowing it to be made public.

Should those who have the direction of our Marine, upon a re-examination of the draught, or out of respect to the opinions of those who have expressed their approbation of it, think proper so far to adopt it as to give it a trial, I cannot help flattering myself that the experiments will turn out of much importance to the public service; and should it answer, as I think there is reason to expect, I shall be amply repaid for my trouble by the satisfaction I shall have in seeing

my endeavours to be of use to my country crowned
with success.

To describe fully the principles upon which this
draught was formed, would be to write a Treatise of
Naval Architecture, which is a work I have not leisure
at present to undertake ; but I would just observe that
my great object was to contrive a vessel, which, possessing all the qualities necessary for a ship of war,
should at the same time be able to carry a great quantity of sail with little ballast.

The *stiffness* of a ship depends upon her form, and
the quantity and stowage of her ballast : but that
vessel which is stiff from construction is much better
adapted for sailing fast than one which, in order to
carry the same quantity of canvas, is obliged to be
loaded with a much greater weight ; for the resistance
is as the quantity of water to be removed, or nearly as
the area of a transverse section of the immersed part
of the body at the midship bend ; and a body that is
broad and shallow is much stiffer than one of the same
capacity that is narrow and deep.

Another advantage attending ships that are stiff
from construction is they are much less liable to
roll than those which are obliged to carry a great
weight of ballast : they are also much better sea-
boats, and are less liable to be strained in bad weather.

Cutters, which are by far the stiffest vessels from
construction of any that have yet been built, are remarkably easy in the sea at all times ; and, I believe,
are safer than any other class of vessels of the same
capacity : they certainly sail faster and work better.

You will see by the draught that I have totally
avoided hollow water-lines, and also that the line of

extreme breadth is everywhere considerably above the
line of flotation. The reasons for this construction you
will immediately comprehend without my mentioning
them, as also many other particulars respecting the
draught, upon which I have not time at present to
enlarge. To the draught, therefore, I shall refer you,
without adding any thing more to this letter, only to
assure you that I really am, etc.

B. THOMPSON.

PALL MALL, March 4, 1781.

PROPOSALS *for Building, by Private Subscription, a*
FRIGATE *upon a new and improved Construction for
Sailing, to be sheathed with Copper, and to carry
Forty Guns and Two Hundred and Fifty Men.*

The essential benefit to the national service which
is attained by every material discovery that directly
leads to naval excellence, and gives a decided superiority at sea, cannot but be an object of the first concern to those who feel for the reputation and safety of
their country, and are anxious for the success and glory
of his Majesty's arms.

The annexed drawing has received the approbation
of some of the best judges of Naval Architecture, both
professional and practical men; who all concur in
opinion, that a ship upon this construction must
necessarily sail much faster than any vessel that has
yet been built; and that, from the manner of arming
her, she will be greatly superior in force to any frigate
in the service.

It is therefore presumed that Naval Architecture
will be brought much nearer perfection by the improve-

ment in the form of this vessel, and a more advantageous system of arming ships of war be introduced, than is at present adopted by any maritime power.

As it may be proper to make some explanation to such professional men as may have these proposals under their eye, of the peculiar construction of this frigate, and of the manner in which it is proposed to arm her, it will be necessary to observe that, to *sail fast* being the great leading principle which governs her whole construction, all the water-lines are perfectly fair, and her body is formed in the most exact and beautiful proportions. This extreme delicacy of form, which is most conspicuous near the keel, will not, however, prevent her giving ample stowage for four months' provisions, besides all her stores; and her great length and breadth above the water will at the same time furnish more commodious room for the men's berths, and better accommodation for the officers, than any frigate in the Navy. Her great length, breadth upon the beam, and good bearings, are qualities that will not only enable her to carry a press of sail, but prevent her rolling and pitching too violently in a rough sea.

It is proposed to give her the masts, yards, and sails of a thirty-two gun frigate, and also the same cables and anchors; and as it sometimes happens in calm weather that very heavy-going ships make their escape from the fastest sailers under favour of light airs, which often extend but to a small distance, to prevent so mortifying an event, and also to enable this frigate to avail herself of any of those favourable opportunities which sometimes occur for attacking ships of force as they lie becalmed, she will be prepared for rowing with thirty oars and one hundred and twenty men, each oar to be twenty-five feet in length, and to be worked by four men. All the oars are to be worked between decks, by running them out at the scuttles that serve occasionally for airing the ship.

Her length upon the main deck being one hundred and fifty feet, it is proposed to pierce her for thirty guns on this deck, and she will carry ten guns upon her quarter-deck, to which may be added two chase-guns upon her forecastle. All the guns upon the main deck are to be thirty-two pounders, upon a new construction, weighing twenty-six hundreds each; and the quarter-deck guns will be light twelve-pounders.

As thirty-two pounder carronades, which are not half so heavy as the proposed thirty-two pounders, have been proved with very large charges of powder, there can be no doubt that these guns may be made to stand fire with perfect safety; and that they will do sufficient execution, and be manageable on shipboard, will appear evident, when it is considered that many of the thirty-two pounders now in use in the Navy weigh no more than fifty-two hundreds, and that they may be fired with two bullets at a time with the greatest possible effect, and without rendering the recoil at all too violent; for it is experimentally true that one bullet may be fired from a gun weighing twenty-six hundreds, with the same velocity, and consequently to the same distance when the elevation is the same, as two fired at once from a piece weighing fifty-two hundreds; and the velocity of the recoil will be the same in both cases.

But, when the velocity of the recoil is the same, the strain upon the breechings will be as the weight of the gun. The force of the recoil, therefore, of these new

PLATE XVI

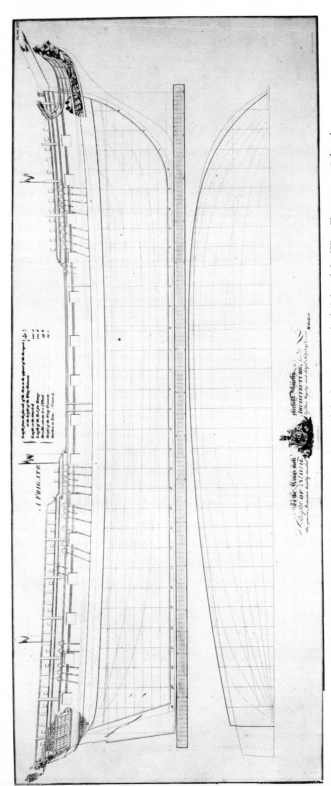

Length from the foreside of the Stem to the aftpart of the Sternpost at the height of the Wing-Transom, 148 ft. 2 in.
Length on the Gun-deck, 150 feet.    Breadth extreme to a 3 in. plank, 39 ft. 6 in.
Length of the Keel for Tonage, 120 ft. 6 in.    Height of the Wing-Transom, 19 ft. 1 in.
Burthen in Tons, No. 1000 5/94.

pieces will be but half as great as that of the thirty-two pounders now in use; and therefore there can be no doubt but they may easily be managed.

The quarter-deck guns are formed upon the same principle, and are just half the weight of the heaviest twelve-pounders in the service.

In order to facilitate the working of the guns, it is proposed to mount them all on sliding carriages, the bed upon which the carriage runs to be movable upon a hinge fastened to the sill of the port in such a manner that the bed may be always kept in a horizontal position, however the ship may lie along, by which means the weather guns may be fought at all times, and the lee guns till their muzzles come down to the water; and that with as much ease and expedition as if the ship was upright upon her keel.

Instead of small arms for the tops, and for the quarter-deck and forecastle, it is proposed to make use of musketoons, on such a construction as to mount on swivel-stocks, and to be used occasionally, either on shipboard or in a boat. These pieces, having a bore of about three feet in length and one inch and a half in diameter, will carry a grape of nine musket-bullets, or eighteen or twenty-four pistol-bullets, as the object is at a greater or less distance, or occasionally a single leaden bullet of twelve ounces, if execution is meant to be done at a very great distance.

PLATE XVII

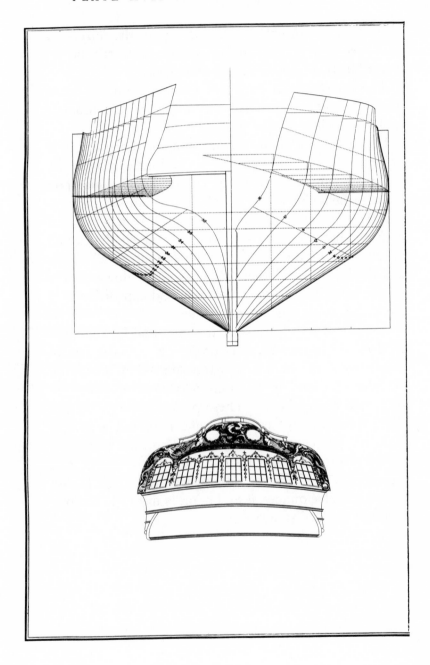

A COMPARATIVE VIEW *of the* DIMENSIONS *of the proposed* FRIGATE *and of the "* LARK *" Frigate of Thirty-two Guns, which was built after a Drawing of the late Mr.* BATELY.

|  | Proposed Frigate. | | The "Lark." | |
|---|---|---|---|---|
|  | Feet. | Inches. | Feet. | Inches. |
| Length of the keel . . . . . | 128 | 0 | 111 | 0 |
| Length on the gun-deck . . . | 150 | 0 | 132 | 0 |
| Extreme breadth . . . . . . | 39 | 6 | 34 | 0 |
| Draught of water, { Forward . . | 15 | 9 | 15 | 6 |
| { Abaft . . . | 15 | 9 | 16 | 6 |
| Area of a transverse section of the immersed part of the body at the midship frame . . . . | 315 | 0 | 378 | 0 |
| Burthen in builder's tonnage . . | 1,000 tons. | | 646 tons. | |
| Real capacity of the immersed part of the body to the load water-line . . . . . . . | 32,784 cubic ft. | | 32,198 cubic ft. | |
| Real burthen . . . . . . . | 915 tons. | | 898½ tons. | |

For the satisfaction of those who may be willing to encourage this undertaking, the following certificates are annexed : —

COPY *of a* LETTER *from* CAPTAIN (*now* REAR-ADMIRAL) KEMPENFELT, *Admiral's Captain in the Grand Fleet.*

[COPY.]

DEAR SIR, — I have viewed the plans for the construction of your intended frigate, and think, as far as I can judge, that she will answer what you expect. Her great length favours the water-lines by diminishing their inflections, and consequently rendering their angles at the extremities more acute. This must greatly facilitate her movement through the water. At the same time, this length of keel, together with the great breadth, will enable her to support much sail, so that from this and the delicacy of her bottom it may be concluded she will go very fast.

The manner you propose to arm this frigate will render her the most formidable, of forty guns, that has yet appeared at sea.

To conclude, you have struck out something new, both for the constructing and arming of a frigate, which in both promises to be a great improvement upon this useful class of vessels. And upon this principle, without taking in other considerations, your proposals merit all encouragement.

I am, with much esteem, dear sir, etc.

RD. KEMPENFELT.

*Charles Street, Westminster*, April 21, 1780.

B. THOMPSON, Esq.

COPY *of a* LETTER *from* SIR CHARLES DOUGLAS, BARONET, *Captain in the Royal Navy, and Commander of his Majesty's Ship* DUKE, *of ninety-eight guns.*

CHARLES STREET, WESTMINSTER, April 23, 1780.

SIR, — I most sincerely acknowledge myself beyond measure obliged to you for having regaled me with the examination of your plan of the frigate of war you propose building; and, having maturely considered the same, I scruple not to give it as my humble opinion that her intended water-lines are better formed for dividing and leaving the fluid than any I have ever yet seen laid down on paper. As also that her general form is such as will insure a requisite degree of stiffness under sail, with far less ballast than ships as they usually are shaped of necessity require, which striking circumstance cannot but be productive of great additional velocity by keeping such part of her body above the water as is the least proper for separating and leaving it, and which must otherwise be immersed; likewise of the desirable effect of carrying her guns higher. Nor have I time sufficiently to expatiate upon these, or to enumerate all the concomitant advantages which I sincerely think the frigate in question will have beyond all such as I have had any knowledge of belonging to this or any other country. I much approve, too, of your ballasting her with iron, with your reprobating the use of shingle for that purpose, and never departing from the general principle of ballasting with the densest attainable matter, ever to be placed as low as possible, that, with less weight thereof than with materials less dense can be effected, the requisite stiffness under sail may be produced, to the great end that the very important purposes mentioned and extensively alluded to in the foregoing may be answered. Upon the whole, then, I do not entertain a doubt of this your proposed frigate sailing with such swiftness as will occasion surprise, nor of her possessing every other eligible quality a ship can have to a most eminent degree. Her force, too, will evidently far exceed that of any ship carrying the same number of men and guns heretofore sent to sea, at least that I have ever seen or heard of. For the sake, then, of the public weal, so much depending upon improvement in our Naval Architecture, may this your plan, so eminently tending thereto, meet with all possible and immediate encouragement; and that you may enjoy perfect health to see the same quickly carried into execution and trial, as also long to enjoy the deserved fruits thereof, is most sincerely and ardently wished by,

Sir, your most, etc.,

CHARLES DOUGLAS.

B. THOMPSON, Esq.

The three first of the following CERTIFICATES are signed by some of the most eminent SHIP-BUILDERS in this KINGDOM, and the last is signed by a gentleman well known in the world as a mathematician.

[COPY.]

I, HAVING seen and examined a draught of a frigate proposed by Mr. Thompson, to be built by private subscription, am of opinion that the said frigate is likely to sail faster than any ship on the present construction in the Navy; and likewise that she promises to be stiff under sail, carry her guns well, and be a good sea-boat. And I think that many advantages will probably be derived to the public from the experiment.

W. WELLS.

LONDON, April 14, 1780.

[COPY.]

I, HAVING seen and examined a draught of a frigate proposed by Mr. Thompson, to be built by private subscription, am of opinion that the said frigate is likely to sail faster than any ship on the present construction in the Navy; and likewise that she promises to be stiff under sail, carry her guns well, and be a good sea-boat. And I think that many advantages will probably be derived to the public from the experiment.

JOHN HALLETT.

LONDON, April 14, 1780.

[COPY.]

HAVING seen and examined the drawing of a frigate upon a new construction, proposed by Mr. Thompson, to be built by subscription, we are of opinion that the said frigate bids fair to sail faster than any vessel that has yet been built; that she will be very stiff under the sail that is proposed to give her, and will be a good sea-boat; that she will carry her guns well out of the water, and, from her great length and breadth upon the gun-deck, will fight them to great advantage. And as it is very probable that many important improvements may be derived to the art of ship-building from the proposed experiment, we think it well worthy of a trial.

W. BARNARD.
JOHN DUDMAN.

LONDON, April 18, 1780.

### [COPY.]

I HAVE examined Mr. Thompson's calculations for determining the capacity of the *Lark* frigate, and of a frigate on a new construction, proposed by him to be built by subscription; and I am of opinion that the capacities of both those frigates are very exactly computed.

CHARLES HUTTON,
*Professor of Mathematics,*
*Royal Military Academy.*

WOOLWICH, April 29, 1780.

[The figures in Plates xvi. and xvii. have been much reduced in size from the original plans.]

# REFERENCES TO RUMFORD'S OWN WORKS

(References are to the present edition)

1. "The Propagation of Heat in Fluids," *Collected Works of Count Rumford*, Vol. I, p. 117.
2. "Of the Use of Steam as a Vehicle for Transporting Heat," Vol. III, p. 1.
3. "Of the Management of Fire and the Economy of Fuel," Vol. II, p. 309.
4. "On the Construction of Kitchen Fire-places and Kitchen Utensils," Vol. III, p. 55.
5. "Of Food; and particularly of feeding the Poor," Vol. IV.
6. "Chimney Fire-places, with Proposals for improving them to save Fuel; to render Dwelling-houses more comfortable and salubrious, and effectually to prevent Chimneys from smoking," Vol. II, p. 221.

# FACTS

# OF

# PUBLICATION

OF THE USE OF STEAM AS A VEHICLE FOR TRANSPORT-
ING HEAT

*Annalen der Physik*, begun by F. A. C. Gren, continued by
L. W. Gilbert (Halle, 1802), XIII, 385–394

Sir Benjamin Thompson, Count of Rumford, *Essays, Political,
Economical and Philosophical* (London: T. Cadell, jr. and W.
Davies, 1802), III, 475–498; Essay XV.

*The Complete Works of Count Rumford* (Boston: American
Academy of Arts and Sciences), II (1873), 324–344.

NOTE ON THE USE OF STEAM HEAT

Read at the Institut de France, June 9, 1806.

*The Complete Works of Count Rumford* (Boston: American
Academy of Arts and Sciences), IV (1875), 789.

This note is translated from the French original, which appears in the *procès verbal* of the Institut de France.

OBSERVATIONS ON THE BEST MEANS OF HEATING THE
HALL IN WHICH THE ORDINARY MEETINGS OF THE
INSTITUTE ARE HELD

Read at the Institut de France, August 10, 1807.

*The Complete Works of Count Rumford* (Boston: American
Academy of Arts and Sciences), IV (1875), 790–795.

This paper is translated from the French original, which appears in the *procès verbal* of the Institut de France for November 23, 1807.

## DESCRIPTION OF A NEW BOILER CONSTRUCTED WITH A VIEW TO THE SAVING OF FUEL

*A Journal of Natural Philosophy, Chemistry and the Arts: Illustrated with Engravings*, edited by William Nicholson (London, 1807), XVII, 5–10 (where this paper is said to have been read at a meeting of the First Class of the French Institute, October 6, 1806). The paper was translated from the French by W. A. Cadell and revised by Count Rumford, from whom it was received.

*Bibliothèque Britannique* (*Science et Arts*), edited by Auguste Pictet, Charles Pictet, and F. G. Maurice (Geneva, 1807), XXXV, 197–205.

*The Complete Works of Count Rumford* (Boston: American Academy of Arts and Sciences), II (1873), 352–358.

## EXPERIMENT ON THE USE OF THE HEAT OF STEAM, IN PLACE OF THAT OF AN OPEN FIRE, IN THE MAKING OF SOAP

*A Journal of Natural Philosophy, Chemistry and the Arts: Illustrated with Engravings*, edited by William Nicholson (London, 1807), XVII, 10–12 (where this paper is said to have been read at a meeting of the First Class of the National Institute, October 20, 1806).

*The Complete Works of Count Rumford* (Boston: American Academy of Arts and Sciences), II (1873), 359–361.

## OF THE MANAGEMENT OF FIRE IN CLOSED FIRE-PLACES

Sir Benjamin Thompson, Count of Rumford, *Essays, Political, Economical and Philosophical* (London: T. Cadell, jr. and W. Davies, 1802), III, 455–471; Essay XIV.

*The Complete Works of Count Rumford* (Boston: American Academy of Arts and Sciences), III (1874), 489–504.

## ON THE CONSTRUCTION OF KITCHEN FIRE-PLACES AND KITCHEN UTENSILS

Sir Benjamin Thompson, Count of Rumford, *Essays, Political, Economical and Philosophical* (London: T. Cadell, jr. and W. Davies, 1802), III, 1–384; Essay X.

Part I, published separately (London: T. Cadell, jr. and W. Davies, 1799).

Part II, published separately (London: T. Cadell, jr. and W. Davies, 1800?).

Part III, published separately (London: T. Cadell, jr. and W. Davies, 1800?).

*The Complete Works of Count Rumford* (Boston: American Academy of Arts and Sciences), III (1874), 167–488.

## OF THE SALUBRITY OF WARM ROOMS

*Bibliothèque Britannique (Science et Arts)*, edited by Auguste Pictet, Charles Pictet, and F. G. Maurice (Geneva, 1802), XX, 119–134.

Sir Benjamin Thompson, Count of Rumford, *Essays, Political Economical and Philosophical* (London: T. Cadell, jr. and W. Davies, 1802), III, 401–417; Essay XII.

*The Complete Works of Count Rumford* (Boston: American Academy of Arts and Sciences), IV (1875), 567–581.

## OF THE SALUBRITY OF WARM BATHING

*Bibliothèque Britannique (Science et Arts)*, edited by Auguste Pictet, Charles Pictet, and F. G. Maurice (Geneva, 1802), XX, 227–249.

Sir Benjamin Thompson, Count of Rumford, *Essays, Political, Economical and Philosophical* (London: T. Cadell, jr. and W. Davies, 1802), III, 419–453; Essay XIII.

*The Complete Works of Count Rumford* (Boston: American Academy of Arts and Sciences), IV (1875), 583–613.

ON THE SPECIFIC GRAVITY, DIAMETER, STRENGTH, COHESION, ETC., OF SILK

Read before the Royal Society, January 11 and 18, 1787.
Royal Society of London, Guard book No. 82, Jan. 18, 1787 to May 17, 1787. Decade IX, No. 23. Letter, 35 pages.
First published by L. M. H. Lowry, C. D. Lowry, Jr., and J. R. Miner, "An Unpublished Paper of Count Rumford," Journal of Chemical Education *9* (1934), 558.
The paper has been transcribed here from the original without correction of spellings, capitalizations, punctuation, or arithmetic.

AN ACCOUNT OF SOME EXPERIMENTS MADE TO DETERMINE THE QUANTITIES OF MOISTURE ABSORBED FROM THE ATMOSPHERE BY VARIOUS SUBSTANCES

Read before the Royal Society, March 22, 1787.
*Philosophical Transactions of the Royal Society of London 77* (London, 1787), 84–124.
Sir Benjamin Thompson, Count of Rumford, *Philosophical Papers* (London: T. Cadell, jr. and W. Davies, 1802), I, 264–269.
*The Complete Works of Count Rumford* (Boston: American Academy of Arts and Sciences), I (1870), 232–236.

EXPERIMENTS AND OBSERVATIONS ON THE ADVANTAGE OF EMPLOYING WHEELS WITH BROAD FELLOES FOR TRAVELLING AND PLEASURE CARRIAGES

Read at a meeting of the First Class of the National Institute, April 15, 1811.
*Le Moniteur Universel ou Gazette Nationale* (Paris, April 25, 1811), 444–446.
Reprinted from *Le Moniteur* in a separate edition (Paris, 1811), pp. 15.

*Bibliothèque Britannique (Science et Arts)*, edited by Auguste Pictet, Charles Pictet, and F. G. Maurice (Geneva, 1811), XLVII, 82–105.

*The Complete Works of Count Rumford* (Boston: American Academy of Arts and Sciences), IV (1875), 661–678.

PLANS FOR THE CONSTRUCTION OF A FRIGATE

Extract from Stalkartt's *Naval Architecture, or the Rudiments and Rules of Ship Building. Exemplified in a Series of Draughts and Plans, with Observations extending to the further Improvement of that Art* (London: J. Boydell, Cheapside, J. Dodsley, Pall Mall, and J. Sewell, Cornhill, 1781; folio, pp. 231).

*The Complete Works of Count Rumford* (Boston: American Academy of Arts and Sciences), IV (1875), 679–691.

# INDEX